Fatty Acids

Seventh supplement to the Fifth Edition of

McCance and Widdowson's

The Composition of Foods

Fatty Acids

Seventh supplement to the Fifth Edition of

McCance and Widdowson's

The Composition of Foods

Compiled by the Ministry of Agriculture, Fisheries and Food

The publishers make no representation, express or implied, with regard to the accuracy of the information contained in this book and cannot accept any legal responsibility or liability for any errors or omissions that may be made.

A catalogue record for this book is available from the British Library.

ISBN 0 85404 819 7

Published by The Royal Society of Chemistry, Cambridge, and the Ministry of Agriculture, Fisheries and Food, London.

Photocomposed by Land & Unwin (Data Sciences) Ltd., Bugbrooke
Printed in the United Kingdom by Redwood Books Ltd., Trowbridge

CONTENTS

Dedicated to the memory of Dr David Buss (1938–1998) who contributed so much to the field of food composition and to *McCance and Widdowson's The Composition of Foods.*

ACKNOWLEDGEMENTS

The Ministry of Agriculture, Fisheries and Food (MAFF) is grateful to the numerous people who have helped during the preparation of this book.

Many current and former professional and administrative staff at MAFF have been involved in the work leading to the production of this book, from design of the analytical projects on which most of the data are based through data collation and checking to the final compilation. In particular, the major contribution of Dr Wynnie Chan is gratefully acknowledged.

Most of the new analyses in this book were undertaken by the Laboratory of Government Chemist. Additional analyses were carried out by Agricultural Development and Advisory Service (ADAS), Aspland & James, the Campden and Chorleywood Food Research Association (CCFRA), Institute of Human Nutrition and Brain Chemistry, Milk Marketing Board (now Milk Marque), Leatherhead Food RA, RHM Technology, and as part of the TRANSFAIR study.

We wish to thank numerous manufacturers, retailers and other organisations for information on the range and composition of their products. In particular, we would like to thank Farley Health Products, Institute of Human Nutrition and Brain Chemistry, Institute of Food Research, Mead Johnsons Nutritionals, Nutricia Ltd (SMA Nutrition and Milupa), Quaker Oats, St Ivels, Van den Bergh Foods, Wyeth and Weetabix for providing additional data.

The preparation of this book was overseen by the Sub Group on Publication of Data, under the auspices of MAFF's Working Party on Nutrients in Food, which comprised Ms Rachael Abraham (Nutrition consultant), the late Dr David Buss (Nutrition consultant), Dr Wynnie Chan (formerly MAFF), Mrs Susan Church (MAFF), Professor Michael Crawford (Institute of Human Nutrition and Brain Chemistry), Miss Alison Paul (MRC Dunn Nutrition Centre), and Professor David Southgate (formerly the Institute of Food Research).

INTRODUCTION

This is the tenth detailed reference book on the nutrients in food, in a series extending and updating the information in McCance and Widdowson's *The Composition of Foods*. It gives the detailed fatty acid composition of a wide selection of foods, including cereals and cereal products, milk and milk products, eggs and egg dishes, fats and oils, meat, poultry and game, meat products and dishes, fish and fish products, vegetables and vegetable dishes, fruit and nuts, preserves, confectionery and snacks, beverages and soups, sauces and dressings.

Interest in the range of fatty acids in foods has grown, and there have been substantial changes in fatty acid composition and in the amounts of fat in foods since previous data on such a wide range of foods was brought together in the fourth Edition of *The Composition of Foods* (Paul & Southgate, 1978) and its first supplement (Paul *et al.*, 1980). With the results of further analyses, the number of foods in this supplement has been increased to 550. There is also an increase in the number of individual fatty acids shown, including, for the first time, positional isomers of the polyunsaturated fatty acids (n-3 and n-6, or ω-3 and ω-6) as well as the total amounts of *cis* and *trans* fatty acids in almost all the foods. (For further explanation of fatty acid nomenclature please see the Appendix on Page 173.)

For ease of use, the values have been given per 100g food (per 100ml feed for infant formulae) rather than per 100g fatty acids. This means that separate values are needed for whole, semi-skimmed and skimmed milks, and for different cuts of beef, for example, because of their differing amounts of fat, even though the proportions of the fatty acids in the fat are almost exactly the same in each. Using information on proportions of the fatty acids, it is possible to calculate the fatty acid composition of almost every food in the UK database, and these proportions have therefore been included in the computerised version of this supplement. For practical reasons, however, the number of foods in this printed supplement has been restricted to major items and ingredients.

This supplement has been compiled by the Ministry of Agriculture, Fisheries and Food. Other supplements in the series, which provide a comprehensive and up-to-date database on nutrients in the wide range of foods now available in Britain, were produced in collaboration with The Royal Society of Chemistry between 1987 and 1996. These covered *Cereals and Cereal Products* (Holland *et al.*,1988), *Milk Products and Eggs* (Holland *et al.*,1989), *Vegetables, Herbs and Spices* (Holland *et al.*, 1991a), *Fruit and Nuts* (Holland *et al.*, 1992a), *Vegetable Dishes* (Holland *et al.*, 1992b), *Fish and Fish Products* (Holland *et al.*, 1993), *Miscellaneous Foods* (Chan *et al.*, 1994), *Meat, Poultry and Game* (Chan *et al.*, 1995); and *Meat Products and Dishes* (Chan *et al.*, 1996). Computer-readable versions of all the supplements are available from HMSO.

Methods

The selection of foods and the determination of fatty acid values follow the general principles used for previous books in this series. The foods were chosen to cover as useful a range of foods as possible of those available and eaten in Britain at the present time. Most of the fatty acids in most of the foods were determined by new analyses, although some values were derived by interpolation, some were obtained from manufacturers and the scientific literature, and a few were taken from the first supplement to *The Composition of Foods* (Paul *et al.*, 1980). It should be noted that the very limited amount of information available on the detailed fatty acid composition of many foods made it particularly difficult to confirm the validity of data.

Literature and manufacturers' values

The scientific literature and manufacturers' information was first reviewed for details of the fatty acid composition of foods. Many of the literature values did not reflect the composition of retail foods as they are sold in Britain today, and most manufacturers' values were for a limited range of fatty acids. Furthermore, few of these were for cooked foods.

Analyses

The majority of the numerous new analyses needed to complete these tables were commissioned by the Ministry of Agriculture, Fisheries and Food from the Laboratory of the Government Chemist (LGC) between 1990 and 1997. Some additional analyses were carried out by RHM Technology; the Agricultural Development and Advisory Service; Aspland & James; the Campden and Chorleywood Food Research Association; Milk Marketing Board (now Milk Marque); Leatherhead Food Research Association; Institute of Human Nutrition and Brain Chemistry; Institute of Food Research; and as part of the EU TRANSFAIR study (van Poppel *et al.*, 1998). Many of the foods included in this supplement and in the computerised version had already been obtained and analysed for other supplements, and were therefore, as described in those supplements, as representative as possible of the items currently bought or eaten in this country. But for many foods, new samples of existing foods were obtained specifically for fatty acids. In addition, there are 139 new foods which have not appeared in previous supplements.

As in previous supplements, individual samples of each raw or cooked food were combined before analysis. The analytical methods were generally as described in the fifth edition of *The Composition of Foods* (Holland *et al.*, 1991b). The individual fatty acids were almost always determined by gas chromatographic analysis, usually of the methyl esters of the fatty acids prepared from the total extracted lipid. They include the fatty acids derived from both triglycerides and phospholipids. Further details of analytical techniques used can be provided on request.

Zeros have sometimes been used in the tables for convenience when fatty acids were not reliably detected or identified in the foods, even though small amounts may still be present. Several other assumptions have been made on the data. For a number of foods where no specific analyses were carried out to measure the proportions of positional isomers, values have mostly been estimated from analysed proportions of similar foods and ascribed in italics. Where no analytical data were available, all 18:2 were assumed to exist in the n-6 position. Similarly, where geometric isomers were not specifically determined, an assumption has been made that for all unprocessed foods, fruit and vegetables and fish (*i.e.* not foods which have undergone hydrogenation, or foods from ruminant animals), mono-unsaturated and polyunsaturated fatty acids exist in the *cis* position. Generally, where positional or geometrical isomerisation is uncertain, values have been italicised.

In addition to fatty acids, the total fat in most food includes the glycerol of the triglycerides, phosphate from phospholipids, and unsaponifiable components such as sterols. To allow the calculation of the total fatty acids in a given weight of food, the conversion factors in Table 1 were applied.

Table 1   Conversion factors to give total fatty acids in fat[a]

Wheat, barley and rye[b]		Beef lean[d]	0.916
whole grain	0.720	Beef fat[d]	0.953
flour	0.670	Lamb, take as beef	
bran	0.820	Pork lean[e]	0.910
		Pork fat[e]	0.953
Oats, whole[b]	0.940	Poultry	0.945
Rice, milled[b]	0.820	Heart[e]	0.789
Milk and milk products	0.945	Kidney[e]	0.747
Eggs[c]	0.830	Liver[e]	0.741
		Fish, fatty[f]	0.900
Fats and oils		white[f]	0.700
all except coconut oil	0.956	Vegetables and fruit	0.800
coconut oil	0.942	Avocado pears	0.956
		Nuts	0.956

[a]Paul & Southgate (1978)
[b]Weihrauch et al. (1976)
[c]Posati et al. (1975)
[d]Anderson et al. (1975)
[e]Anderson (1976)
[f]Exler et al. (1975)

A worked example of the calculation is given below (TFA = total fatty acids):

Total fat in Beef, lean only   =   5.1g/100g
Conversion factor   =   0.916
Total fatty acids in food = total fat x conversion factor = 5.1 x 0.916 = 4.7g/100g

Saturates	at 43.7g/100g	TFA x 4.7 ÷ 100 =	2.0g/100g food
cis-Monounsaturates	at 47.9g/100g	TFA x 4.7 ÷ 100 =	2.2g/100g food
cis-Polyunsaturates	at 3.8g/100g	TFA x 4.7 ÷ 100 =	0.2g/100g food

N.B. The values do not add up to the total fatty acids because branched-chain and trans fatty acids have been excluded from the saturated and unsaturated acids respectively.

Arrangement of the tables

Food groups

For ease of reference, the foods in this book have been listed alphabetically within the following groups: cereals and cereal products; milk and milk products; eggs and egg dishes; fats and oils; meat, poultry and game, meat products and dishes; fish and fish products; vegetables and vegetable dishes; fruit and nuts; preserves, confectionery and snacks; beverages and soups; sauces and dressings. Where foods or products are cooked, values for the raw food are given first. Values for dishes made up of complex ingredients have not been included in this supplement because of the substantial variations which are possible. In general, foods with less than 1g fat per 100g food have been excluded from this printed supplement unless they are important sources of fat or major ingredients in foods.

Numbering system

For ease of reference, each food has been assigned a consecutive publication number for the purposes of this supplement only. In addition, each food has a unique food code number which is given in the index and will allow read-across to other supplements, where appropriate. For foods that have already been included in supplements or in the 5th edition, their food code number (including the unique 2 digit prefix) has been repeated. These prefixes are 11 – *Cereals and Cereal Products*; 12 – *Milk Products and Eggs*; 13 – *Vegetables, Herbs and Spices*; 14 – *Fruit and Nuts*; 15 – *Vegetable Dishes*; 16 – *Fish and Fish Products*; 17 – *Miscellaneous Foods*; 18 – *Meat, Poultry and Game*; 19 – *Meat Products and Dishes*; 50 – *Fifth Edition*. Foods that have not previously been included such as commercial versions of earlier recipes have been given a new food code number in the supplement using that prefix *e.g.* beef stir-fried with green peppers and black bean sauce (19-305). Where a new set of samples was analysed for fatty acids and the fat content differed from the original samples, a new code number has been given but with the same supplement prefix, *e.g.* short sweet biscuits were 11-184, now 11-382. For ease of use, the original food code number is given alongside the new one in the index for the foods concerned. These are the numbers that will be used in nutrient databank applications.

Description and number of samples

The information given under this heading includes the source and number of samples taken for analysis, or the source of any literature or manufacturers' values.

Fatty acids

The values for each food are shown on four consecutive pages. All values are given as grams per 100 grams of the food as described, but values in grams per 100g fatty acids can be found in the computerised version of the supplement.

In addition to the description for each food, the first page presents a broad picture of the fat with the total amount of fat and, wherever possible, the total amounts of *cis* monounsaturated fatty acids, *cis* polyunsaturated fatty acids, total n-3 and total n-6 polyunsaturated fatty acids separately, total *trans* fatty acids (monounsaturates and polyunsaturates together), and branched-chain fatty acids.

The second page for each food shows the amounts of the main individual saturated fatty acids. The third page covers the monounsaturated fatty acids, with the amounts of the individual *cis* isomers, and where possible, columns for important positional isomers (*cis* and *trans*) and for *trans* monounsaturated fatty acids. The fourth page for each food shows the amounts of individual *cis* polyunsaturated fatty acids (where possible) arranged in n-6 and n-3 series, and for *trans* polyunsaturated fatty acids.

Appendices

There are 3 appendices in this supplement. The first shows the common and systematic names for the fatty acids in this supplement. The second and third appendices give the cholesterol and phytosterol content of foods which have not previously appeared in the fifth Edition of *The Composition of Foods* or in any of the supplements.

Nutrient variability

Almost all foods vary in nutritional value, and this can be particularly important for the fatty acid composition of many foods. There are two sources of variation. The first is that the amount of fat in the food may vary, either naturally, as with seasonal differences in fatty fish, or because manufacturers vary the amounts or quality of the ingredients in their products. Varying amounts of fat may also be removed from some foods, particularly meat, before or at the retail level, or by caterers or in the home either before or after cooking. This fat may be trimmed off, or varying amounts may be lost (or gained in frying) during the cooking itself. Users of these tables should ensure that the food they are considering matches as far as possible the description given in the tables, or allow for any differences in the amount of fat if this is known from the food label.

The fatty acid composition of the fat in the food may also vary, for similar reasons. Because of the range of possible fat sources for many manufactured foods, depending in part upon price and availability and the flexibility given by processes such as hydrogenation, there can be major changes in the fatty acid composition of some manufactured foods with no change in the product name or description. But if the changes have been made to meet, for example, new health considerations, some details may be given on the label. A range of fats of different fatty acid composition can also be used when foods are cooked at home. This supplement therefore excludes most domestic recipe dishes, but users may wish to calculate their own values where the amounts and type of fat are known.

The values in this supplement are as representative as possible of the foods at the time of analysis. Users requiring details of possible recent changes in the fatty acid composition of specific fats and oils may wish to contact the manufacturer directly.

The introduction to the fifth edition of *The Composition of Foods* gives more detailed accounts of these and other factors that should be taken into account in the proper use of food composition tables. Users of the present supplement are advised to read them and take them to heart.

References to introductory text

Anderson, B.A., Kinsella, J.A. and Watt, B.K (1975) Comprehensive evaluation of fatty acids in foods. II. Beef products. *J. Am. Diet. Assoc.* **67**: 35-41

Anderson, B.A. (1976) Comprehensive evaluation of fatty acids in foods . VII. Pork products. *J. Am. Diet. Assoc.* **69**: 44-49

Chan, W., Brown, J. and Buss, D.H. (1994) *Miscellaneous Foods*. Fourth supplement to 5th edition of *McCance and Widdowson's The Composition of Foods*. The Royal Society of Chemistry, Cambridge

Chan, W., Brown, J., Church, S.M. and Buss, D.H. (1996) *Meat Products and Dishes*. Sixth supplement to 5th edition of *McCance and Widdowson's The Composition of Foods*. The Royal Society of Chemistry, Cambridge

Chan, W., Brown, J., Lee, S.M. and Buss, D.H. (1995) *Meat, Poultry and Game*. Fifth supplement to 5th edition of *McCance and Widdowson's The Composition of Foods*. The Royal Society of Chemistry, Cambridge

Exler, J., Kinsella, J.E. and Watt, B.K. (1975) Lipids and fatty acids of important finfish. New data for nutrient tables. *J. Am. Oil Chem. Soc.* **52**: 154-159

Holland, B., Brown, J. and Buss, D.H. (1993) *Fish and Fish products*. Third supplement to 5th edition of *McCance and Widdowson's The Composition of Foods*. The Royal Society of Chemistry, Cambridge

Holland, B., Unwin, I.D. and Buss, D.H. (1988) *Cereals and Cereal Products*. Third supplement to *McCance and Widdowson's The Composition of Foods*. The Royal Society of Chemistry, Cambridge

Holland, B., Unwin, I.D. and Buss, D.H. (1989) *Milk Products and Eggs*. Fourth supplement to *McCance and Widdowson's The Composition of Foods*. The Royal Society of Chemistry, Cambridge

Holland, B., Unwin, I.D. and Buss, D.H. (1991a) *Vegetables, Herbs and Spices*. Fifth supplement to *McCance and Widdowson's The Composition of Foods*. The Royal Society of Chemistry, Cambridge

Holland, B., Unwin, I.D. and Buss, D.H. (1992a) *Fruit and Nuts*. First supplement to 5th edition of *McCance and Widdowson's The Composition of Foods*. The Royal Society of Chemistry, Cambridge

Holland, B., Welch, A.A., Unwin, I.D., Buss, D.H., Paul, A.A. and Southgate, D.A.T. (1991b) *McCance and Widdowson's The Composition of Foods*, 5th edition. The Royal Society of Chemistry, Cambridge

Holland, B., Welch, A.A. and Buss, D.H. (1992b) *Vegetable Dishes*. Second supplement to 5th edition of *McCance and Widdowson's The Composition of Foods*. The Royal Society of Chemistry, Cambridge

Paul, A.A. and Southgate, D.A.T. (1978) *McCance and Widdowson's The Composition of Foods*, 4th edition. HMSO, London

Paul, A.A., Southgate, D.A.T. and Russell, J. (1980) First supplement *to McCance and Widdowson's The Composition of Foods,* 4th edition: *Amino acid composition (mg per 100g food) and fatty acid composition (g per 100g food)*. HMSO, London

Posati, L.P., Kinsella, J.E. and Watt, B.K. (1975) Comprehensive evaluation of fatty acids in foods. III. Eggs and egg products. *J. Am. Diet. Assoc.* **67**: 111-115

van Poppel, G., van Erp-Baart, M , I eth, T., Gevers, E., Van Amelsvoort, J., Lanzmann-Petithory, D., Kafatos, A., Aro, A. (1998) *Trans* fatty acids in foods in Europe: the TRANSFAIR study. *Journal of Food Composition and Analysis*, **11**: 112-136

Weihrauch, J.L., Kinsella, J.E. and Watt, B.K. (1976) Comprehensive evaluation of fatty acids in foods. VI. Cereal products. *J. Am. Diet. Assoc.* **68**: 335-340

The Tables

Symbols and abbreviations used in the tables

Symbols

0	None of the nutrient is present, or when fatty acids were not reliably detected or identified in the food
Tr	Trace. Used where the fatty acid was detected but below levels judged to be quantifiable.
N	The nutrient is present in significant quantities but there is no reliable information on the amount
Italics	Estimated value

Abbreviations

Satd	Saturated
Monounsatd	Monounsaturated
Polyunsatd	Polyunsaturated

Cereals and cereal products

Fat and total fatty acids, g per 100g food

No.	Food	Description	Total fat	Satd	cis-Mono unsatd	Polyunsatd Total cis	n-6	n-3	Total trans	Total branched
	Grains and flours									
1	**Barley**, pearl, *raw*	2 samples from different shops	1.7	0.29	0.14	0.77	0.70	0.07	0	0
2	**Oatmeal**, quick cook, *raw*	10 samples, 8 brands	9.2	1.61	3.34	3.71	3.52	0.19	0	0
3	**Wheat flour**, brown	Representative samples from Voluntary Flour Sampling Scheme	2.0	0.27	0.22	0.85	0.80	0.05	0	0
4	white, household, *plain*	Representative samples from Voluntary Flour Sampling Scheme	1.2	0.16	0.13	0.51	0.48	0.03	0	0
5	wholemeal	Representative samples from Voluntary Flour Sampling Scheme	2.2	0.28	0.21	0.89	0.83	0.06	0	0
6	**Rye flour**, whole	Analytical and literature sources	2.0	0.27	0.21	0.95	0.82	0.13	0	0
7	**Soya flour**, *full fat*	Mixed sample	23.5	3.30	5.70	13.34	11.68	1.66	0	0
8	*reduced fat*	Mixed sample	7.2	1.00	1.74	4.09	3.58	0.51	0	0
	Rice and pasta									
9	**Rice**, brown, *raw*	5 assorted samples	2.8	0.74	0.66	0.98	0.94	0.04	0	0
10	**Rice**, white, easy cook, *raw*	10 samples, 9 brands; parboiled	3.6	0.85	0.91	1.29	1.26	0.03	0	0
11	**Pasta**, plain, fresh, *cooked* [a]	12 samples, 8 brands including tagliatelle, spaghetti, lasagna, linguine and fusilli	1.5	0.28	0.28	0.36	0.34	0.02	0.01	0
	Breads									
12	**Bread**, white, average	Average of 5 main types of white sliced and unsliced bread	1.9	0.40	0.25	0.66	0.62	0.04	0	0

[a] Contains 0.03g unidentified fatty acids per 100g food

Cereals and cereal products

Saturated fatty acids, g per 100g food

No.	Food	4:0	6:0	8:0	10:0	12:0	14:0	15:0	16:0	17:0	18:0	20:0	22:0	24:0
	Grains and flours													
1	**Barley**, pearl, *raw*	0	0	0	0	0	Tr	0	0.29	0	Tr	0	0	0
2	**Oatmeal**, quick cook, *raw*	0	0	0	0	0	0	0	1.47	0	0.10	0.04	0	0
3	**Wheat flour**, brown	0	0	0	0	0	0	0	0.25	0	0.01	0.01	0	0
4	white, household, *plain*	0	0	0	0	0	0	0	0.15	0	0.01	0.01	0	0
5	wholemeal	0	0	0	0	0	0	0	0.25	0	0.01	0.01	0	0
6	**Rye flour**, whole	0	0	0	0	0	0	0	0.27	0	Tr	0	0	0
7	**Soya flour**, *full fat*	0	0	0	0	0.02	0.04	0	2.25	0	0.90	0.07	0.02	0
8	*reduced fat*	0	0	0	0	Tr	0.01	0	0.69	0	0.28	0.02	Tr	0
	Rice and pasta													
9	**Rice**, brown, *raw*	0	0	0	0	0	0.04	0	0.65	0	0.05	0	0	0
10	**Rice**, white, easy cook, *raw*	0	0	0	0	0	0.03	0	0.73	0	0.08	0.01	0	0
11	**Pasta**, plain, fresh, *cooked*	0	0	0	0	0	0	0	0.24	0	0.05	0	0	0
	Breads													
12	**Bread**, white, average	0	0	0	0	Tr	0.02	0	0.31	0	0.07	0	0	0

Cereals and cereal products

Monounsaturated fatty acids, g per 100g food

No.	Food	cis 10:1	12:1	14:1	15:1	16:1	17:1	18:1	cis/trans 18:1 n-9	cis/trans 18:1 n-7	cis 20:1	22:1	cis/trans 22:1 n-11	22:1 n-9	cis 24:1	trans Monounsatd
Grains and flours																
1	**Barley**, pearl, *raw*	0	0	0	0	Tr	0	0.14	0.14	0	0	0	0	0	0	0
2	**Oatmeal**, quick cook, *raw*	0	0	0	0	0.02	0	3.32	3.32	0	0	0	0	0	0	0
3	**Wheat flour**, brown	0	0	0	0	0.01	0	0.21	0.21	0	0.01	0	0	0	0	0
4	white, household, *plain*	0	0	0	0	Tr	0	0.12	0.12	0	Tr	0	0	0	0	0
5	wholemeal	0	0	0	0	Tr	0	0.21	0.21	0	Tr	0	0	0	0	0
6	**Rye flour**, whole	0	0	0	0	Tr	0	0.20	0.20	0	0.01	0	0	0	0	0
7	**Soya flour**, *full fat*	0	0	0	0	0.04	0	5.62	5.62	0	0.04	0	0	0	0	0
8	*reduced fat*	0	0	0	0	0.01	0	1.72	1.72	0	0.01	0	0	0	0	0
Rice and pasta																
9	**Rice**, brown, *raw*	0	0	0	0	0.01	0	0.65	0.65	0	0	0	0	0	0	0
10	**Rice**, white, easy cook, *raw*	0	0	0	0	Tr	0	0.91	0.91	0	0	0	0	0	0	0
11	**Pasta**, plain, fresh, *cooked*	0	0	0	0	0.01	0	0.27	0.27	0	0	0	0	0	0	0.01
Breads																
12	**Bread**, white, average	0	0	0	0	0.01	0	0.24	0.24	0	0	0	0	0	0	0

Cereals and cereal products

Polyunsaturated fatty acids, g per 100g food

No.	Food	cis n-6					cis n-3					trans
		18:2	18:3	20:3	20:4	22:4	18:3	18:4	20:5	22:5	22:6	Polyunsatd
Grains and flours												
1	**Barley**, pearl, *raw*	0.70	0	0	0	0	0.07	0	0	0	0	0
2	**Oatmeal**, quick cook, *raw*	3.52	0	0	0	0	0.19	0	0	0	0	0
3	**Wheat flour**, brown	0.80	0	0	0	0	0.05	0	0	0	0	0
4	white, household, *plain*	0.48	0	0	0	0	0.03	0	0	0	0	0
5	wholemeal	0.83	0	0	Tr	0	0.06	0	0	0	0	0
6	**Rye flour**, whole	0.82	0	0	0	0	0.13	0	0	0	0	0
7	**Soya flour**, *full fat*	11.68	0	0	0	0	1.66	0	0	0	0	0
8	*reduced fat*	3.58	0	0	0	0	0.51	0	0	0	0	0
Rice and pasta												
9	**Rice**, brown, *raw*	0.94	0	0	0	0	0.04	0	0	0	0	0
10	**Rice**, white, easy cook, *raw*	1.26	0	0	0	0	0.03	0	0	0	0	0
11	**Pasta**, plain, fresh, *cooked*	0.34	0	0	0	0	0.02	0	0	0	0	0
Breads												
12	**Bread**, white, average	0.62	0	0	0	0	0.04	0	0	0	0	0

Cereals and cereal products

Fat and total fatty acids, g per 100g food

No.	Food	Description	Total fat	Satd	cis-Mono unsatd	Polyunsatd Total cis	n-6	n-3	Total trans	Total branched
	Breads continued									
13	**Bread**, wholemeal, average	Average of 3 types of sliced and unsliced wholemeal bread	2.9	0.54	0.41	1.16	1.08	0.08	0	0
14	softgrain	10 samples, 5 brands	1.4	0.34	0.25	0.52	0.47	0.05	0.01	0
15	speciality, white	9 samples including ciabatta, focaccia and pugliese	4.2	0.67	2.04	0.86	0.80	0.06	0	0
	Rolls									
16	**Croissants**, plain, retail [a]	Data from Transfair study; 10 samples from assorted outlets	26.0	14.33	6.62	1.17	1.00	0.41	1.64	0.16
17	savoury, retail	4 samples from assorted outlets	23.5	9.95	6.92	1.89	1.84	0.25	3.71	N
18	sweet, retail	4 samples from assorted outlets	18.3	11.28	4.48	0.58	0.52	0.12	0.86	0.30
	Breakfast cereals									
19	**All-Bran**	3 samples of the same brand	3.5	0.57	0.45	1.82	1.71	0.14	0.02	0
20	**Bran Flakes**	Data from Institute of Human Nutrition and Brain Chemistry	1.9	0.35	0.18	1.03	0.96	0.07	Tr	0
21	**Cornflakes**	Data from Institute of Human Nutrition and Brain Chemistry	0.7	0.24	0.12	0.25	0.19	0.06	Tr	0
22	**Crunchy Nut Corn Flakes**	3 samples of the same brand	3.5	0.42	1.57	0.89	0.89	Tr	Tr	0
23	**Crunchy Oat Cereal**	3 samples of the same brand	15.9	4.13	4.74	2.42	2.40	0.05	3.60	0
24	**Optima/Fruit 'n Fibre**	3 samples of the same brand	4.6	1.96	0.88	0.43	0.43	0.03	0.04	0

a Contains 0.91g unidentified fatty acids per 100g food

Cereals and cereal products

Saturated fatty acids, g per 100g food

No.	Food	4:0	6:0	8:0	10:0	12:0	14:0	15:0	16:0	17:0	18:0	20:0	22:0	24:0
	Breads *continued*													
13	**Bread**, wholemeal, average	0	0	0	0	0	0.04	0	0.41	0	0.09	0	0	0
14	softgrain	0	0	0	0	0	0	0	0.27	0	0.05	0	0	0
15	speciality, white	0	0	0	0	0	0	0	0.55	0	0.10	0.01	0.01	0
	Rolls													
16	**Croissants**, plain, retail	N	N	0.40	0.60	0.69	2.50	0.54	6.20	0.22	3.13	0.04	0.01	0
17	savoury, retail	0.09	0.07	0.07	0.09	0.52	0.56	0.04	6.04	0.04	2.31	0.09	0.02	0
18	sweet, retail	0.86	0.47	0.26	0.54	0.66	2.05	0.21	4.29	0.10	1.82	0.02	Tr	0
	Breakfast cereals													
19	**All-Bran**	0	0	0	0	Tr	Tr	Tr	0.52	0.01	0.03	0	0	0
20	**Bran Flakes**	0	0	0	0	0	0	0	0.35	0	0	0	0	0
21	**Cornflakes**	0	0	0	0	0	0	0	0.22	0	0.01	0.01	0	0
22	**Crunchy Nut Corn Flakes**	0	0	0	0	0	Tr	0	0.34	0	0.08	0	0	0
23	**Crunchy Oat Cereal**	0	0.01	0.12	0.11	0.95	0.40	0.01	1.98	0.01	0.55	0	0	0
24	**Optima/Fruit 'n Fibre**	0	0.01	0.11	0.10	0.92	0.38	0	0.35	0	0.09	0	0	0

No.	Food	10:1	12:1	14:1	cis 15:1	cis 16:1	17:1	18:1	cis/trans 18:1 n-9	18:1 n-7	cis 20:1	22:1	cis/trans 22:1 n-11	22:1 n-9	cis 24:1	trans Monounsatd
Breads continued																
13	**Bread**, wholemeal, average	0	0	0	0	0.04	0	0.37	0.37	0	0	0	0	0	0	0
14	softgrain	0	0	0	0	Tr	Tr	0.24	0.23	0.02	0.01	0	0	0	0	Tr
15	speciality, white	0	0	0	0	0.03	0	2.00	2.00	0	0.01	0	0	0	0	0
Rolls																
16	**Croissants**, plain, retail	0	0	0	0	0.29	0	6.12	4.33	1.41	0.21	0	0	0	0	1.40
17	savoury, retail	0	0	0.02	0	0.07	0	6.78	9.12	0.88	0.04	0	0	0	0	3.50
18	sweet, retail	0.05	0	0.17	0.05	0.24	0.05	3.88	4.22	0.17	0.02	0	0	0	0	0.79
Breakfast cereals																
19	**All-Bran**	0	0	0	0	0.01	0	0.45	0.45	0	0	0	0	0	0	Tr
20	**Bran Flakes**	0	0	0	0	0	0	0.18	0.18	0	0	0	0	0	0	Tr
21	**Cornflakes**	0	0	0	0	0	0	0.12	0.12	0	0	0	0	0	0	Tr
22	**Crunchy Nut Corn Flakes**	0	0	0	0	0	0	1.57	1.57	0	0	0	0	0	0	Tr
23	**Crunchy Oat Cereal**	0	0	0	0	0.02	0	4.72	8.30	0	0	0	0	0	0	3.58
24	**Optima/Fruit 'n Fibre**	0	0	0	0	Tr	0	0.87	0.89	0	0	0	0	0	0	0.01

Cereals and cereal products

Polyunsaturated fatty acids, g per 100g food

No.	Food	cis n-6					cis n-3					trans
		18:2	18:3	20:3	20:4	22:4	18:3	18:4	20:5	22:5	22:6	Polyunsatd
Breads continued												
13	**Bread**, wholemeal, average	1.08	0	0	0	0	0.08	0	0	0	0	0
14	softgrain	0.47	0	0	0	0	0.05	0	0	0	0	Tr
15	speciality, white [a]	0.80	0	0	0	0	0.06	0	0	0	0	0
Rolls												
16	**Croissants**, plain, retail [b]	0.78	0	0.01	0	0	0.37	0	0	0	0	0.24
17	savoury, retail	1.64	0	0	0	0	0.25	0	0	0	0	0.20
18	sweet, retail	0.45	0	0	0	0	0.12	0	0	0	0	0.07
Breakfast cereals												
19	**All-Bran**	1.68	0	0	0.02	0	0.12	0	0	0	0	0.02
20	**Bran Flakes**	0.96	0	0	0	0	0.07	0	0	0	0	Tr
21	**Cornflakes**	0.19	0	0	0	0	0.06	0	0	0	0	Tr
22	**Crunchy Nut Corn Flakes**	0.89	0	0	0	0	0	0	0	0	0	Tr
23	**Crunchy Oat Cereal**	2.35	0	0	0.05	0	0.03	0	0	0	0	0.02
24	**Optima/Fruit 'n Fibre**	0.43	0	0	0	0	0	0	0	0	0	0.03

a Contains 0.01g 20:2 per 100g food
b Contains 0.01g 20:2 per 100g food

Cereals and cereal products

Fat and total fatty acids, g per 100g food

No.	Food	Description	Total fat	Satd	cis-Mono unsatd	Polyunsatd Total cis	n-6	n-3	Total trans	Total branched
	Breakfast cereals continued									
25	**Muesli**	3 samples of the same brand; contains wheatflakes, oatflakes, sultanas, dates, hazelnuts, raisins, apple and coconut	6.7	1.19	3.35	1.77	1.69	0.05	0.04	0
26	**Puffed Wheat**	Manufacturer's data (Quaker)	1.3	0.20	0.16	0.58	0.54	0.04	Tr	0
27	**Ready Brek**	Data from Institute of Human Nutrition and Brain Chemistry	7.8	1.36	2.83	3.14	2.98	0.16	Tr	0
28	**Shredded Wheat**	3 samples of the same brand	2.2	0.34	0.26	0.97	0.90	0.07	0.01	0
29	**Sugar Puffs**	Manufacturer's data (Quaker)	1.0	0.15	0.12	0.44	0.41	0.03	Tr	0
30	**Weetabix**	3 samples of the same brand	2.0	0.29	0.22	0.92	0.85	0.07	Tr	0
	Biscuits									
31	**Cheese sandwich biscuits**	10 samples, 5 brands	35.3	23.66	7.45	1.72	1.75	0.06	0.72	0.12
32	**Chocolate biscuits**, full coated	7 samples, 5 brands including Breakaway, United and chocolate fingers	25.3	13.83	6.68	1.13	1.24	0.10	2.37	0.05
33	cream filled, full coated	9 samples of different brands including Club, Penguin, Tric and Hob Nob bars	28.4	16.29	7.87	1.57	N	N	N	0.05
34	**Chocolate chip cookies**	16 samples, 8 brands	22.9	10.64	8.38	2.51	2.44	0.12	0.28	0.03
35	**Cornish wafers**	8 samples of the same brand	28.3	21.18	4.65	1.14	N	N	N	N
36	**Cream crackers**	10 samples of the same brand	13.3	5.42	4.59	1.40	1.38	0.05	1.24	N
37	**Crunch biscuits**, cream filled	5 samples, 2 brands of crunch creams	24.6	15.00	5.93	1.69	1.67	0.09	0.89	N
38	**Digestives**, chocolate, half coated	22 samples, 4 brands; plain and milk	24.1	12.22	7.57	1.28	1.51	0.08	1.60	0.10

Saturated fatty acids, g per 100g food

No.	Food	4:0	6:0	8:0	10:0	12:0	14:0	15:0	16:0	17:0	18:0	20:0	22:0	24:0
	Breakfast cereals continued													
25	**Muesli**	0	0	0.03	0.02	0.21	0.09	0	0.72	0	0.12	0	0	0
26	**Puffed Wheat**	0	0	Tr	Tr	Tr	Tr	Tr	0.17	Tr	0.01	Tr	Tr	Tr
27	**Ready Brek**	0	0	0	0	0	0	0	1.24	0	0.08	0.04	0	0
28	**Shredded Wheat**	0	0	Tr	Tr	Tr	0.01	Tr	0.30	Tr	0.02	0	0	0
29	**Sugar Puffs**	0	0	0	Tr	Tr	Tr	Tr	0.14	0	0.01	Tr	Tr	Tr
30	**Weetabix**	0	0	0	0	Tr	Tr	Tr	0.27	0	0.02	0	0	0
	Biscuits													
31	**Cheese sandwich biscuits**	0.14	0.08	0.66	0.72	7.63	3.62	0.06	8.30	0.06	2.38	0	0	0
32	**Chocolate biscuits,** full coated	0.06	0.04	0.03	0.07	0.11	0.40	0.05	7.20	0.06	5.69	0.10	0.02	0
33	cream filled, full coated	0.05	0.05	0.24	0.24	2.42	1.14	0.05	6.90	0.05	4.94	0.16	0.03	0
34	**Chocolate chip cookies**	0.03	0.02	0.02	0.03	0.13	0.38	0.02	7.69	0.05	2.25	Tr	0	0
35	**Cornish wafers**	0	0.04	0.69	0.74	11.99	4.27	N	2.79	0.01	0.65	0	0	0
36	**Cream crackers**	0	0	Tr	Tr	0.04	0.11	0.01	4.47	0.01	0.77	0	0	0
37	**Crunch biscuits,** cream filled	0.02	0.05	0.42	0.38	4.09	1.62	Tr	6.44	0.02	1.86	0.07	0.02	0
38	**Digestives,** chocolate, half coated	0.05	0.02	0.02	0.03	0.11	0.60	0.04	7.47	0.05	3.57	0.16	0.12	0

Cereals and cereal products

Monounsaturated fatty acids, g per 100g food

No.	Food	cis							cis/trans		cis		cis/trans		cis	trans
		10:1	12:1	14:1	15:1	16:1	17:1	18:1	18:1 n-9	18:1 n-7	20:1	22:1	22:1 n-11	22:1 n-9	24:1	Monounsatd
	Breakfast cereals continued															
25	**Muesli**	0	0	0	0	0.01	0	3.34	3.38	0	0	0	0	0	0	0.04
26	**Puffed Wheat**	0	0	0	0	Tr	0	0.14	0.14	0	0.01	Tr	Tr	0	0	N
27	**Ready Brek**	0	0	0	0	0.02	0	2.81	2.81	0	0	0	0	0	0	Tr
28	**Shredded Wheat**	0	0	0	0	Tr	0	0.26	0.26	0	0	0	0	0	0	Tr
29	**Sugar Puffs**	0	0	0	0	Tr	0	0.11	0.11	0	0.01	Tr	Tr	Tr	0	N
30	**Weetabix**	0	0	0	0	0	0	0.22	0.22	0	0	0	0	0	0	Tr
	Biscuits															
31	**Cheese sandwich biscuits**	0.01	0	0.04	0	0.10	0.02	7.29	7.34	0.39	0	0	0	0	0	0.62
32	**Chocolate biscuits**, full coated a	Tr	0	0.03	0	0.09	0.02	6.53	4.83	3.80	0.01	0	0	0	0	2.18
33	cream filled, full coated	Tr	Tr	0.03	0	0.08	0.03	7.71	7.03	1.76	0.03	0.03	0	0.03	0	1.30
34	**Chocolate chip cookies** b	Tr	0	0.01	0	0.06	0.01	8.30	8.37	0.15	0	0	0	0	0	0.23
35	**Cornish wafers**	0	0	0	0	0	N	4.65	4.18	0.42	0	0	0	0	0	0.05
36	**Cream crackers**	0	0	0	0	0.03	0.01	4.56	5.13	0.52	0	0	0	0	0	1.21
37	**Crunch biscuits**, cream filled	0	0	0	0	0.02	Tr	5.88	6.07	0.42	0.02	0	0	0	0	0.87
38	**Digestives**, chocolate, half coated	0	0	0.02	0.01	0.15	0.02	7.16	6.19	1.32	0.10	0.11	0.02	0.25	0	1.29

a Contains 0.05g other monounsaturated fatty acids per 100g food
b Contains 0.01g other monounsaturated fatty acids per 100g food

Cereals and cereal products

Polyunsaturated fatty acids, g per 100g food

No.	Food	cis n-6					cis n-3					trans
		18:2	18:3	20:3	20:4	22:4	18:3	18:4	20:5	22:5	22:6	Polyunsatd
Breakfast cereals *continued*												
25	**Muesli**	1.69	0	0	0.03	0	0.05	0	0	0	0	Tr
26	**Puffed Wheat**	*0.54*	0	0	0	0	*0.04*	0	Tr	0	Tr	Tr
27	**Ready Brek**	2.98	0	0	0	0	0.16	0	0	0	0	Tr
28	**Shredded Wheat**	0.90	0	0	0.01	0	0.06	0	0	0	0	0.01
29	**Sugar Puffs**	0.41	0	0	0	0	0.03	0	0	0	Tr	Tr
30	**Weetabix**	0.85	0	0	0.01	0	0.07	0	0	0	0	Tr
Biscuits												
31	**Cheese sandwich biscuits**	*1.68*	0	0	0	0	*0.04*	0	0	0	0	0.09
32	**Chocolate biscuits**, full coated	*1.07*	0	0	0	0	0.08	0	0	0	0	0.19
33	cream filled, full coated	1.47	0	0	0	0	0.11	0	0	0	0	N
34	**Chocolate chip cookies**	*2.40*	0	0	0	0	0.12	0	0	0	0	0.05
35	**Cornish wafers**	1.12	0	0	0	0	0.02	0	0	0	0	N
36	**Cream crackers**	*1.36*	0	0	0	0	0.04	0	0	0	0	0.03
37	**Crunch biscuits**, cream filled	*1.65*	0	0	0	0	0.05	0	0	0	0	0.02
38	**Digestives**, chocolate, half coated	*1.22*	0	0	0	0	0.07	0	0	0	0	0.31

Fat and total fatty acids, g per 100g food

No.	Food	Description	Total fat	Satd	cis-Mono unsatd	Polyunsatd Total cis	n-6	n-3	Total trans	Total branched
Biscuits continued										
39	**Digestives**, plain	10 samples, 4 brands	20.3	9.00	7.37	1.89	1.86	0.09	0.95	0.06
40	**Fig rolls**	10 samples, 5 brands	6.2	2.81	2.30	0.76	0.69	0.09	0.02	0.01
41	**Fruit biscuits**	7 samples, 4 brands including Jaspers, Shrewsburys and fruit shortcakes	19.7	8.31	6.97	2.07	N	N	N	0.04
42	**Gingernut biscuits**	10 samples, 5 brands	13.0	6.04	4.59	1.20	1.23	0.03	0.53	0.02
43	**Jaffa cakes**	10 samples, 4 brands	10.1	5.41	3.37	0.73	0.68	0.06	N	0.01
44	**Krackerwheat**	5 samples of the same brand	30.0	18.88	6.66	1.95	2.00	0.06	1.09	tr
45	**Oatcakes**	8 samples, 3 brands	15.1	5.07	6.00	2.83	2.67	0.18	0.37	N
46	**Oat based biscuits**	10 samples, 3 brands including Hob Nobs, Snapjacks, Oatbakes and Barnstormers	21.4	9.23	7.55	2.31	2.41	0.06	0.93	0.11
47	chocolate, half coated	20 samples, 4 brands including Hob Nobs, Barnstormers and Oatbakes	23.8	11.03	8.18	2.04	2.07	0.06	1.21	0.06
48	**Sandwich biscuits**, cream filled	20 samples, 5 brands including custard creams and bourbon	20.7	11.03	5.48	1.64	1.28	0.57	2.03	0.04
49	jam filled	6 samples, 3 brands including Jammy Dodgers and jam rings	17.3	7.19	6.12	1.90	1.87	0.13	1.32	N
50	**Semi-sweet biscuits**	10 samples, 3 brands including Osborne, Rich Tea and Marie	13.3	6.25	4.37	1.18	1.22	0.04	0.80	0.02
51	**Shortcake**, chocolate, half coated	6 samples of the same brand including Shorties, Animals, Signature and Magic Numbers	21.8	11.48	6.81	1.19	1.13	0.10	1.10	0.17

Cereals and cereal products

No.	Food	4:0	6:0	8:0	10:0	12:0	14:0	15:0	16:0	17:0	18:0	20:0	22:0	24:0
	Biscuits continued													
39	**Digestives**, plain	0	0	N	Tr	0	0.32	0.03	7.32	0.04	1.28	0	0	0
40	**Fig rolls**	0	0	Tr	Tr	0.02	0.07	0.01	2.39	0.01	0.31	0	0	0
41	**Fruit biscuits**	0	0	Tr	Tr	0.06	0.43	0.02	6.44	0.04	1.05	0.15	0.11	0
42	**Gingernut biscuits**	0	0	Tr	0.01	0.06	0.19	0.01	5.01	0.02	0.73	0	0	0
43	**Jaffa cakes**	0.01	Tr	Tr	0.01	0.01	0.05	0.01	2.35	N	2.97	0	0	0
44	**Krackerwheat**	0	0.05	0.73	0.69	6.13	2.69	0.02	7.13	0.02	1.42	0	0	0
45	**Oatcakes**	0	0	0.01	Tr	0.04	0.11	0.01	4.18	0.01	0.71	0	0	0
46	**Oat based biscuits**	0	0	Tr	0.01	0.05	0.33	0.02	7.35	0.05	1.41	0	0	0
47	**Oat based biscuits**, chocolate, half coated	0.01	0.01	0.01	0.02	0.05	0.20	0.02	7.26	N	3.48	0	0	0
48	**Sandwich biscuits**, cream filled	0	0.01	0.23	0.22	2.42	1.07	0.01	5.49	0.02	1.56	N	N	0
49	jam filled	0	0	0.02	0.02	0.10	0.17	Tr	5.94	0.02	0.88	0.05	0.02	0
50	**Semi-sweet biscuits**	0	0	0.01	0.01	0.16	0.28	0.02	4.90	0.03	0.85	N	N	0
51	**Shortcake**, chocolate, half coated	0.21	0.15	0.08	0.19	0.25	0.83	0.10	5.34	0.08	4.13	0.10	0.02	0

23

Cereals and cereal products

Monounsaturated fatty acids, g per 100g food

No.	Food	10:1	12:1	cis 14:1	15:1	16:1	17:1	18:1	cis/trans 18:1 n-9	18:1 n-7	20:1	cis 22:1	cis/trans 22:1 n-11	22:1 n-9	cis 24:1	trans Monounsatd
	Biscuits continued															
39	**Digestives**, plain	0	0	0.01	0	0.18	0.02	7.16	7.16	0.72	0	0	0	0	0	0.89
40	**Fig rolls**	0	0	0	0	0.01	N	2.29	2.04	0.21	0	0	0	0	0	Tr
41	**Fruit biscuits**	0	0	0	0	0.09	0.02	6.20	5.56	1.49	0.21	0.45	0.19	0	0	1.45
42	**Gingernut biscuits**	0	0	0	0	0.06	Tr	4.53	4.45	0.45	0	0	0	0	0	0.47
43	**Jaffa cakes**	0	0	0	0	0.04	Tr	3.32	2.95	0.30	0	0	0	0	0	N
44	**Krackerwheat**	0	0	0	0	0.06	Tr	6.60	6.74	0.69	0	0	0	0	0	0.98
45	**Oatcakes**	0	0	0	0	0.03	Tr	5.96	5.61	0.57	0	0	0	0	0	0.34
46	**Oat based biscuits**	0	0	0	0	0.09	0.01	7.45	7.32	0.74	0	0	0	0	0	0.77
47	chocolate, half coated	0	0	0	0	0.05	0.01	8.12	8.22	0.83	0	0	0	0	0	1.12
48	**Sandwich biscuits**, cream filled	0	0	0	0	0.03	Tr	5.45	6.11	0.76	0	0	0	0	0	1.82
49	jam filled	0	0	0	0	0.02	Tr	6.07	5.61	1.70	0.03	0	0	0	0	1.31
50	**Semi-sweet biscuits**	0	0	0	0	0.13	0.01	4.23	4.25	0.49	0	0	0	0	0	0.72
51	**Shortcake**, chocolate, half coated	0.02	0	0.06	0	0.13	0.04	6.52	6.56	0.67	0.04	0	0	0	0	1.06

Polyunsaturated fatty acids, g per 100g food

No.	Food	cis n-6					cis n-3					trans
		18:2	18:3	20:3	20:4	22:4	18:3	18:4	20:5	22:5	22:6	Polyunsatd
Biscuits *continued*												
39	**Digestives**, plain	1.80	0	0	0	0	0.08	0	0	0	0	0.06
40	**Fig rolls**	0.67	0	0	0	0	0.09	0	0	0	0	0.02
41	**Fruit biscuits** [a]	1.92	0	0	0	0	0.11	0	0	0	0	N
42	**Gingernut biscuits**	1.17	0	0	0	0	0.03	0	0	0	0	0.06
43	**Jaffa cakes**	0.67	0	0	0	0	0.06	0	0	0	0	0.01
44	**Krackerwheat**	1.89	0	0	0	0	0.06	0	0	0	0	0.11
45	**Oatcakes**	2.66	0	0	0	0	0.17	0	0	0	0	0.03
46	**Oat based biscuits**	2.26	0	0	0	0	0.05	0	0	0	0	0.16
47	chocolate, half coated	1.99	0	0	0	0	0.05	0	0	0	0	0.09
48	**Sandwich biscuits**, cream filled	1.08	0	0	0	0	0.57	0	0	0	0	0.21
49	jam filled	1.85	0	0	0	0	0.05	0	0	0	0	0.02
50	**Semi-sweet biscuits**	1.15	0	0	0	0	0.03	0	0	0	0	0.08
51	**Shortcake**, chocolate, half coated	1.08	0	0	0	0	0.10	0	0	0	0	0.04

[a] Contains 0.04g 20:2 per 100g food

Cereals and cereal products

Fat and total fatty acids, g per 100g food

No.	Food	Description	Total fat	Satd	cis-Mono unsatd	Polyunsatd Total cis	n-6	n-3	Total trans	Total branched
Biscuits *continued*										
52	**Short sweet biscuits**	10 samples, 2 brands including Lincoln and shortcake	21.8	11.05	7.38	1.28	1.45	0.05	0.90	0.07
53	**Wafers**, filled, chocolate, full coated	11 samples, different brands including Taxi and Blue Riband	29.7	18.24	7.37	0.99	N	N	N	0.06
54	**Water biscuits**	10 samples, 2 brands	7.0	4.40	1.26	0.82	0.80	0.04	0.19	Tr
Cakes										
55	**Chocolate cup cake**	2 samples, 2 brands	5.1	2.22	1.94	0.65	N	N	N	N
56	**Chocolate covered marshmallow teacake**	10 samples, 5 brands	16.3	8.93	5.30	1.09	1.00	0.12	0.25	N
57	**Fancy iced cakes**	8 samples, 3 brands including French and fondant fancies	9.1	3.21	2.85	0.54	0.60	0.12	2.04	0.05
58	**Fruit cake**, plain	10 samples, 3 brands; sultana	14.8	6.89	4.69	1.08	1.00	0.16	1.34	0.07
59	rich	10 samples, 4 brands	8.2	4.13	2.42	0.59	0.58	0.08	0.32	0.22
60	**Gâteau**, chocolate based	10 samples including Black Forest gâteau	15.7	8.99	3.58	1.00	1.04	0.17	0.39	0.41
61	fruit, frozen	10 samples; fruit and cream sponge including strawberry, orange and lemon and tropical fruit	12.3	7.04	2.80	0.78	0.81	0.13	0.30	0.32
62	**Madeira cake**	10 samples including lemon	15.1	8.43	3.28	1.39	1.34	0.27	0.71	0.29
63	**Sponge cake**, butter cream filling[a]	Data from TRANSFAIR study; 8 samples from different outlets	16.7	7.39	5.15	2.52	2.30	0.31	0.51	0.06
64	**Swiss roll**, chocolate	8 samples, 4 brands	16.8	7.02	5.24	1.22	1.26	0.22	2.23	0.21

a Contains 0.28g unidentified fatty acids per 100g food

Saturated fatty acids, g per 100g food

No.	Food	4:0	6:0	8:0	10:0	12:0	14:0	15:0	16:0	17:0	18:0	20:0	22:0	24:0
	Biscuits continued													
52	**Short sweet biscuits**	0.01	0.01	0.01	0.01	0.05	0.31	0.03	7.09	0.05	3.48	N	N	0
53	**Wafers,** filled, chocolate, full coated	0.08	0.08	0.40	0.40	3.81	1.64	0.03	5.42	0.06	6.13	0.17	0.03	0
54	**Water biscuits**	0	0	0.01	0.16	1.52	0.69	0.07	1.70	Tr	0.25	0	0	0
	Cakes													
55	**Chocolate cup cake**	0	0	0	0	0.01	0.07	0	1.67	0.00	0.45	0.01	0	0
56	**Chocolate covered marshmallow teacake**	0	0.03	0.02	0.05	0.08	0.23	0.03	4.36	0.05	3.90	0.14	0.03	0.02
57	**Fancy iced cakes**	0	0.02	0.02	0.03	0.33	0.25	0.01	1.67	0.03	0.86	0	0	0
58	**Fruit cake,** plain	0.07	0.05	0.04	0.12	0.19	0.77	0.08	4.05	0.11	1.40	0	0	0
59	rich	0.10	0.03	0.02	0.06	0.20	0.76	0.08	2.64	0.05	0.19	0	0	0
60	**Gâteau,** chocolate based	0	0.23	0.23	0.42	0.80	1.46	0.14	3.95	0.10	1.57	0.04	0.04	0.02
61	fruit, frozen	0	0.18	0.18	0.33	0.62	1.15	0.11	3.09	0.08	1.23	0.03	0.03	0.01
62	**Madeira cake**	0.14	0.05	0.04	0.11	0.18	1.36	0.12	4.91	0	1.51	0	0	0
63	**Sponge cake,** butter cream, filling	N	N	0.09	0.17	0.26	0.73	0.14	4.66	0.10	1.17	0.06	0.01	0
64	**Swiss roll,** chocolate	0.05	0.03	0.05	0.09	0.31	0.76	0.07	4.06	0.07	1.51	0	0	0

Cereals and cereal products

Monounsaturated fatty acids, g per 100g food

No.	Food	10:1	12:1	14:1	15:1	cis 16:1	17:1	18:1	cis/trans 18:1 n-9	cis/trans 18:1 n-7	cis 20:1	cis 22:1	cis/trans 22:1 n-11	cis/trans 22:1 n-9	cis 24:1	trans Monounsatd
	Biscuits continued															
52	**Short sweet biscuits**	0	0	0.01	0	0.17	0.01	7.19	6.64	0.86	0	0	0	0	0	0.68
53	**Wafers**, filled, chocolate, full coated	0	0	0.03	0	0.08	Tr	7.14	6.44	2.12	0.03	0	0	0	0	1.64
54	**Water biscuits**	0	0	0	0	0.01	Tr	1.25	1.27	0.13	0	0	0	0	0	0.17
	Cakes															
55	**Chocolate cup cake**	0	0	0	0	0.04	N	1.89	1.90	0.05	0.01	0	0	0	0	0.12
56	**Chocolate covered marshmallow teacake**	0	0	0	0	0.06	0.02	5.20	5.08	0	0.02	0	0	0	0	0.22
57	**Fancy iced cakes**	0	0	0	0	0.19	0.01	2.65	2.65	0	0	0	0	0	0	1.86
58	**Fruit cake**, plain	0.01	0	0.05	0	0.21	0.02	4.40	4.40	0	0	0	0	0	0	1.25
59	rich	0.01	0	0.05	0	0.19	0.02	2.16	2.16	0	0	0	0	0	0	0.25
60	**Gâteau**, chocolate based	0	0	0	0	0.19	0.04	3.33	3.33	0	0.02	0	0	0	0	0.18
61	fruit, frozen	0	0	0	0	0.15	0.03	2.61	2.61	N	0.02	0	0	0	0	0.14
62	**Madeira cake**	0	0	0	0	0.96	0.02	2.29	2.29	0	0	0	0	0	0	0.49
63	**Sponge cake**, butter cream, filling	0	0	0	0	0.13	0	4.98	4.05	0.74	0.04	0	0	0	0	0.43
64	**Swiss roll**, chocolate	0.01	0	0.03	0	0.35	0.02	4.84	4.84	0	0	0	0	0	0	1.97

Cereals and cereal products

Polyunsaturated fatty acids, g per 100g food

No.	Food	cis n-6					cis n-3					trans
		18:2	18:3	20:3	20:4	22:4	18:3	18:4	20:5	22:5	22:6	Polyunsatd
Biscuits continued												
52	**Short sweet biscuits**	1.24	0	0	0	0	0.04	0	0	0	0	0.22
53	**Wafers**, filled, chocolate, full coated	0.93	0	0	0	0	0.06	0	0	0	0	N
54	**Water biscuits**	0.79	0	0	0	0	0.04	0	0	0	0	0.01
Cakes												
55	**Chocolate cup cake**	0.55	0	0	0	0	0.10	0	0	0	0	N
56	**Chocolate covered marshmallow teacake**	0.98	0	0	0	0	0.11	0	0	0	0	0.03
57	**Fancy iced cakes**	0.45	0	0	0	0	0.09	0	0	0	0	0.18
58	**Fruit cake**, plain	0.95	0	0	0	0	0.13	0	0	0	0	0.08
59	rich	0.53	0	0	0	0	0.06	0	0	0	0	0.07
60	**Gâteau**, chocolate based	0.85	0	0	0	0	0.15	0	0	0	0	0.21
61	fruit, frozen	0.66	0	0	0	0	0.12	0	0	0	0	0.16
62	**Madeira cake**	1.21	0	0	0	0	0.18	0	0	0	0	0.22
63	**Sponge cake**, butter cream filling	2.23	0	0	0	0	0.30	0	0	0	0	0.08
64	**Swiss roll**, chocolate	1.06	0	0	0	0	0.16	0	0	0	0	0.26

Cereals and cereal products

Fat and total fatty acids, g per 100g food

No.	Food	Description	Total fat	Satd	cis-Mono unsatd	Polyunsatd Total cis	n-6	n-3	Total trans	Total branched
Cakes *continued*										
65	**Torte**, frozen/chilled, fruit	8 samples, 5 brands including lemon, raspberry, and passion fruit	15.5	8.49	3.88	0.70	0.82	0.14	0.57	0.40
Pastry										
66	**Puff pastry**, frozen, *raw* [a]	Data from TRANSFAIR study; 6 samples from different outlets	24.5	11.65	8.30	3.06	2.61	0.55	0.21	0.01
67	**Shortcrust pastry**, frozen, *raw* [b]	Data from TRANSFAIR study; 6 samples from different outlets	28.5	11.40	11.00	3.97	3.24	0.83	1.07	0.02
Buns and pastries										
68	**Custard tarts**, individual	8 samples, 4 brands	14.5	6.10	5.01	1.35	1.25	0.18	1.08	0.10
69	**Danish pastries**	8 samples of different varieties	14.1	8.57	1.54	1.51	1.65	0.24	0.84	0.42
70	**Doughnuts**, ring	8 samples from different bakeries	22.4	5.81	8.62	5.76	5.46	0.62	1.10	N
71	**Jam tarts**, retail	10 samples, 3 brands; assorted jam	14.7	6.64	4.21	1.64	1.49	0.27	1.04	0.51
72	**Scones**, retail	Data from TRANSFAIR study; 12 samples from different outlets	13.4	3.92	4.80	1.74	1.54	0.38	2.11	0.01
73	**Teacakes**, retail [c]	Data from TRANSFAIR study; 15 samples from different cutlets, including hot cross buns	5.0	1.93	1.73	0.96	0.82	0.16	0.10	0.01

[a] Contains 0.19g unidentified fatty acids per 100g food
[b] Contains 0.21g unidentified fatty acids per 100g food
[c] Contains 0.05g unidentified fatty acids per 100g food

Saturated fatty acids, g per 100g food

No.	Food	4:0	6:0	8:0	10:0	12:0	14:0	15:0	16:0	17:0	18:0	20:0	22:0	24:0
	Cakes continued													
65	**Torte**, frozen/chilled, fruit	0	0.21	0.15	0.35	0.44	1.35	0.15	4.15	0.10	1.52	0.04	0.02	0.01
	Pastry													
66	**Puff pastry**, frozen, *raw*	0	0	0.13	0.03	0.03	0.24	0.03	9.82	0.05	1.15	0.12	0.04	0
67	**Shortcrust pastry**, frozen, *raw*	0	0	0.15	0.04	0.05	0.20	0.03	9.25	0.05	1.45	0.14	0.05	0.01
	Buns and pastries													
68	**Custard tarts**, individual	0.02	0.01	0.01	0.03	0.05	0.58	0.05	4.06	0.05	1.21	0	0	0
69	**Danish pastries**	0	0	N	0.01	0.05	0.57	0.05	5.28	0.10	2.50	0	0	0
70	**Doughnuts**, ring	0	0	Tr	0	0.03	0.13	0.01	4.49	0.03	1.13	0	0	0
71	**Jam tarts**, retail	0	0	N	0	0.04	1.35	0.05	3.98	0.08	1.13	0	0	0
72	**Scones**, retail	0	0	0.08	0.03	0.06	0.29	0.03	2.32	0.04	0.94	0.10	0.05	0
73	**Teacakes**, retail	0	0	0.05	0.05	0.04	0.13	0.02	1.28	0.01	0.30	0.02	0.01	0

Monounsaturated fatty acids, g per 100g food

No. Food	cis							cis/trans		cis		cis/trans		cis	trans
	10:1	12:1	14:1	15:1	16:1	17:1	18:1	18:1 n-9	18:1 n-7	20:1	22:1	22:1 n-11	22:1 n-9	24:1	Monounsatd
Cakes *continued*															
65 **Torte**, frozen/chilled, fruit	0	0	0	0	0.18	0.04	3.64	3.64	0	0.02	0.01	Tr	Tr	0	0.31
Pastry															
66 **Puff pastry**, frozen, *raw*	0	0	0	0	0.05	0	8.11	7.16	0.90	0.13	0.01	Tr	Tr	0	0.11
67 **Shortcrust pastry**, frozen, *raw*	0	0	0	0.03	0.06	0	10.74	9.21	1.41	0.15	0.02	N	N	0.01	0.97
Buns and pastries															
68 **Custard tarts**, individual	0	0	0.02	0	0.35	0.01	4.64	4.64	0	0	0	0	0	0	0.99
69 **Danish pastries**	0.01	0	0.02	0	0.24	0.04	1.23	1.23	0	0	0	0	0	0	0.47
70 **Doughnuts**, ring	0	0	0	0	0.08	0.01	8.53	8.53	0	0	0	0	0	0	0.79
71 **Jam tarts**, retail	0	0	0	0	0.22	0.01	3.99	3.99	0	0	0	0	0	0	0.92
72 **Scones**, retail	0	0	0	0	0.14	0	4.50	3.08	1.43	0.13	0.02	N	N	0	1.93
73 **Teacakes**, retail	0	0	0	0	0.02	0	1.67	1.30	0.38	0.03	0.01	Tr	Tr	0	0.08

No.	Food	18:2	cis n-6				18:3	cis n-3				trans
			18:3	20:3	20:4	22:4		18:4	20:5	22:5	22:6	Polyunsatd
Cakes continued												
65	**Torte**, frozen/chilled, fruit	0.58	0	0	0	0	0.12	0	0	0	0	0.26
Pastry												
66	**Puff pastry**, frozen, *raw*	2.55	0	0	0	0	0.50	0	0	0	0	0.10
67	**Shortcrust pastry**, frozen, *raw*	3.18	0	0	0	0	0.78	0	0	0	0	0.10
Buns and pastries												
68	**Custard tarts**, individual	1.19	0	0	0	0	0.16	0	0	0	0	0.09
69	**Danish pastries**	1.30	0	0	0	0	0.21	0	0	0	0	0.37
70	**Doughnuts**, ring	5.19	0	0	0	0	0.58	0	0	0	0	0.31
71	**Jam tarts**, retail [a]	1.42	0	0	0	0	0.22	0	0	0	0	0.12
72	**Scones**, retail	1.41	0	0.04	0	0	0.27	0	0.01	0	0	0.18
73	**Teacakes**, retail	0.80	0	0	0	0	0.15	0	0	0	0	0.02

[a] Contains 0.01g 20:3 per 100g food

Cereals and cereal products

Fat and total fatty acids, g per 100g food

No.	Food	Description	Total fat	Satd	cis-Mono unsatd	Polyunsatd Total cis	n-6	n-3	Total trans	Total branched
Puddings										
74	**Bakewell tart**, individual	10 samples, 3 brands; cherry	18.4	9.86	4.48	0.90	0.92	0.11	0.88	0.88
75	**Fruit pies**, individual, double crust	10 samples, 3 brands; apple	14.0	5.36	5.09	1.56	1.42	0.26	1.20	0.18
76	**Rice desserts**, with fruit, individual, chilled	7 samples, 2 brands; apple, strawberry, and raspberry. Rice pudding dessert with puréed fruit	2.2	1.45	0.50	0.09	0.08	0.01	0.04	0
77	**Rice pudding**, canned	10 samples, 2 brands	1.3	0.77	0.30	0.06	0.06	0.01	0.03	0.03
78	**Sponge pudding**, canned	8 samples, 4 brands; treacle and golden syrup	9.1	4.98	2.45	0.39	0.46	0.04	0.65	0.13
Savouries										
79	**Cheese nachos**, takeaway	10 samples from different outlets, nachos with melted cheese	18.1	6.69	7.00	3.02	2.41	0.67	0.17	0
80	**Pizza**, thin base, cheese and tomato, takeaway	10 samples from different outlets	10.3	4.75	2.92	1.16	1.10	0.22	0.39	0.20
81	-, fish topped, takeaway	10 samples from different outlets. Toppings include tuna and prawns	8.0	3.24	2.27	1.19	1.13	0.19	0.34	0.13
82	deep pan, cheese and tomato, takeaway	10 samples from different outlets	7.5	3.06	2.22	1.23	1.10	0.24	0.24	0.13
83	-, meat topped, takeaway	8 samples from different outlets. Toppings include pepperoni, beef and pork	9.0	3.45	3.10	1.23	1.11	0.24	0.25	0.13

Cereals and cereal products

Saturated fatty acids, g per 100g food

No.	Food	4:0	6:0	8:0	10:0	12:0	14:0	15:0	16:0	17:0	18:0	20:0	22:0	24:0
Puddings
74	**Bakewell tart**, individual	0	0	0.04	0.05	0.78	1.70	0.11	5.00	0.14	2.03	0	0	0
75	**Fruit pies**, individual, double crust	0	0	Tr	Tr	0.04	0.31	0.03	3.97	0.03	0.97	0	0	0
76	**Rice desserts**, with fruit, individual, chilled	0.08	0.04	0.02	0.05	0.07	0.22	0.02	0.70	0.01	0.22	0.01	0	0
77	**Rice pudding**, canned	0.04	0.02	0.01	0.03	0.04	0.12	0.01	0.33	0.01	0.15	0	0	0
78	**Sponge pudding**, canned	0.07	0.04	0.03	0.09	0.14	0.78	0.07	2.76	0.05	0.96	0	0	0
Savouries
79	**Cheese nachos**, takeaway	0	0.10	0.08	0.19	0.27	0.90	0.11	3.47	0.08	1.37	0.06	0.02	0.02
80	**Pizza**, thin base, cheese and tomato, takeaway	N	0.15	0.08	0.19	0.23	0.71	0.08	2.38	0.06	0.82	0.03	0.02	0.01
81	–, fish topped, takeaway	N	0.09	0.05	0.11	0.14	0.46	0.05	1.67	0.04	0.57	0.02	0.02	0.01
82	deep pan, cheese and tomato, takeaway	N	0.09	0.05	0.11	0.14	0.44	0.05	1.56	0.04	0.53	0.02	0.01	0.02
83	–, meat topped, takeaway	N	0.08	0.04	0.10	0.12	0.41	0.05	1.83	0.05	0.72	0.03	0.01	0.01

Cereals and cereal products

Monounsaturated fatty acids, g per 100g food

No.	Food	cis 10:1	12:1	14:1	15:1	16:1	17:1	18:1	cis/trans 18:1 n-9	18:1 n-7	cis 20:1	22:1	cis/trans 22:1 n-11	22:1 n-9	cis 24:1	trans Monounsatd
Puddings																
74	**Bakewell tart**, individual	0	0	0⁻	0	0.98	0	3.49	3.49	0	0	0	0	0	0	0.94
75	**Fruit pies**, individual, double crust	0	0	Tr	0	0.22	0.01	4.87	4.87	0	0	0	0	0	0	0.93
76	**Rice desserts**, with fruit, individual, chilled	0.01	0	0.02	0	0.04	0	0.43	0.43	0	0	0	0	0	0	0.04
77	**Rice pudding**, canned	0	0	0	0	0.02	0	0.29	0.29	0	0	0	0	0	0	0.01
78	**Sponge pudding**, canned	0.01	0	0.03	0	0.39	0.02	2.00	2.52	0	0	0	0	0	0	0.52
Savouries																
79	**Cheese nachos**, takeaway	0.02	0.02	0.08	0	0.21	0.04	6.49	N	N	0.13	0.01	0	0.01	0	0.14
80	**Pizza**, thin base, cheese and tomato, takeaway	0	0	0	0	0.11	0.02	2.74	2.74	0	0.03	0.01	0	0.01	0	0.22
81	-, fish topped, takeaway	0	0	0	0	0.10	0.02	2.04	2.04	0	0.09	0.01	0	0.01	0	0.20
82	deep pan, cheese and tomato, takeaway	0	0	0	0	0.08	0.02	2.08	2.08	0	0.04	0.01	0	0.01	0	0.13
83	-, meat topped, takeaway	0	0	0	0	0.14	0.03	2.86	2.86	0	0.06	0.01	0	0.01	0	0.13

Cereals and cereal products

Polyunsaturated fatty acids, g per 100g food

No.	Food	cis n-6					cis n-3					trans
		18:2	18:3	20:3	20:4	22:4	18:3	18:4	20:5	22:5	22:6	Polyunsatd
Puddings												
74	**Bakewell tart**, individual	0.92	0	0	0	0	0.11	0	0	0	0	0.13
75	**Fruit pies**, individual, double crust	1.35	0	0	0	0	0.21	0	0	0	0	0.27
76	**Rice desserts**, with fruit, individual, chilled	0.08	0	0	0	0	0.01	0	0	0	0	0
77	**Rice pudding**, canned [a]	0.05	0	0	0	0	0.01	0	0	0	0	0.02
78	**Sponge pudding**, canned	0.35	0	0	0	0	0.04	0	0	0	0	0.13
Savouries												
79	**Cheese nachos**, takeaway	2.33	0.04	0.02	0.01	0.01	0.56	0.01	0.05	0.04	0	0.02
80	**Pizza**, thin base, cheese and tomato, takeaway	0.96	0	0	0	0	0.19	0	0	0	0	0.16
81	–, fish topped, takeaway [b]	1.03	0	0	0	0	0.16	0	0	0	0	0.13
82	deep pan, cheese and tomato, takeaway [c]	1.01	0	0	0	0	0.21	0	0	0	0	0.11
83	–, meat topped, takeaway	1.02	0	0	0	0	0.21	0	0	0	0	0.12

[a] Contains 0.02g 22 poly per 100g food
[b] Contains 0.12g 20 poly, 0.10g 22 poly per 100g food
[c] Contains 0.04g 20 poly, 0.01g 22 poly per 100g food

Milk and milk products

Fat and total fatty acids, g per 100g food

No.	Food	Description	Total fat	Satd	cis-Mono unsatd	Polyunsatd Total cis	n-6	n-3	Total trans	Total branched
Cows' milk										
84	**Skimmed milk**, pasteurised, average	Average of summer and winter milk	0.3	0.13	0.06	0.01	Tr	Tr	Tr	Tr
85	**Semi-skimmed milk**, pasteurised, average [a]	Average of summer and winter milk	1.7	1.07	0.39	0.04	0.05	0.01	0.07	0.03
86	**Whole milk**, pasteurised, average [b]	Average of summer and winter milk	4.0	2.48	0.93	0.10	0.10	0.02	0.14	0.07
87	sterilised	6 samples, 3 brands	3.9	2.38	0.92	0.12	0.07	0.03	0.14	0.08
88	UHT	Average of 22 samples of summer and winter milk	3.9	2.36	0.93	0.12	0.07	0.04	0.15	0.09
89	**Breakfast milk**, pasteurised, average	Average of 6 samples of summer and winter milk	4.7	3.02	1.00	0.13	0.08	0.03	0.15	0.09
90	**Organic semi-skimmed milk**, pasteurised	6 samples, 3 brands	1.8	1.15	0.39	0.05	0.04	0.01	0.05	0.04
91	**Evaporated milk**, light	7 samples, 4 brands	4.1	2.50	0.96	0.11	0.07	0.04	0.17	0.09
Other milks, dairy products and substitutes										
92	**Coffee compliment** powder	10 samples, own brands	21.2	19.63	0.22	0.04	0.02	0.02	0.08	0
93	low fat	8 samples, 3 brands	16.1	14.81	0.24	0.04	0.04	0.01	0.22	0
94	**Coffee whitener** liquid, with skimmed milk and non milk fat	4 samples, 2 brands	4.0	3.53	0.14	0.02	0.02	0	0.13	0
95	with glucose syrup and vegetable fat	4 samples, 2 brands	14.4	13.16	0.26	0.02	0.02	0	0.29	0
96	**Chocolate flavoured milk**, pasteurised	9 samples, 6 brands	0.9	0.58	0.16	0.02	0.02	0.01	0.07	0
97	**Flavoured milk**, pasteurised	10 samples, 6 brands; strawberry, banana	1.5	0.99	0.31	0.04	0.03	0.01	0.04	0
98	**Goats' milk**	6 samples, 5 brands; pasteurised and unpasteurised	3.7	2.37	0.83	0.13	0.13	0.03	0.14	0.01

a Contains 0.06g unidentified fatty acids per 100g food
b Contains 0.13g unidentified fatty acids per 100g food

Milk and milk products

Saturated fatty acids, g per 100g food

No.	Food	4:0	6:0	8:0	10:0	12:0	14:0	15:0	16:0	17:0	18:0	20:0	22:0	24:0
	Cows' milk													
84	**Skimmed milk**, pasteurised, average	0.01	Tr	Tr	0.01	0.01	0.02	Tr	0.06	Tr	0.03	0	0	0
85	**Semi-skimmed milk**, pasteurised, average	0.06	0.04	0.02	0.05	0.07	0.18	0.02	0.46	0.01	0.18	0	0	0
86	**Whole milk**, pasteurised, average	0.14	0.09	0.05	0.11	0.15	0.41	0.04	1.06	0.02	0.41	0	0	0
87	sterilised	0.16	0.09	0.05	0.10	0.12	0.38	0.04	1.00	0.03	0.41	0.01	0	0
88	UHT	0.15	0.09	0.05	0.10	0.12	0.37	0.04	0.98	0.03	0.43	0.01	0	0
89	**Breakfast milk**, pasteurised, average	0.19	0.11	0.06	0.13	0.15	0.48	0.05	1.30	0.03	0.50	0.01	0	0
90	**Organic semi-skimmed milk**, pasteurised	0.07	0.04	0.02	0.05	0.07	0.19	0.02	0.49	0.01	0.17	0	0	0
91	**Evaporated milk**, light	0.17	0.09	0.05	0.11	0.13	0.40	0.04	1.04	0.03	0.42	0.01	0	0
	Other milks, dairy products and substitutes													
92	**Coffee compliment** powder	0	0.10	1.36	1.08	8.61	3.47	0	2.20	0.02	2.72	0.04	0	0.02
93	low fat	0	0.08	0.94	0.68	6.04	2.07	0	1.52	0	3.41	0.04	0.01	0.01
94	**Coffee whitener liquid**, with skimmed milk and non milk fat	0	0.02	0.22	0.18	1.56	0.60	0	0.43	0	0.51	0.01	0	0
95	with glucose syrup and vegetable fat	0	0.08	0.89	0.68	5.81	2.33	0	1.32	0	2.02	0.03	0	0.01
96	**Chocolate flavoured milk**, pasteurised	0.03	0.02	0.01	0.02	0.03	0.08	0.01	0.26	0.01	0.13	0	0	0
97	**Flavoured milk**, pasteurised	0.06	0.03	0.02	0.04	0.05	0.16	0.02	0.43	0.01	0.16	0	0	0
98	**Goats' milk**	0.07	0.08	0.09	0.29	0.16	0.35	0.05	0.93	0.02	0.31	0.01	0.01	0

Milk and milk products

Monounsaturated fatty acids, g per 100g food

No.	Food	cis						18:1	cis/trans 18:1 n-9	cis/trans 18:1 n-7	cis 20:1	cis 22:1	cis/trans 22:1 n-11	cis/trans 22:1 n-9	cis 24:1	trans Monounsatd
		10:1	12:1	14:1	15:1	16:1	17:1	18:1	18:1 n-9	18:1 n-7	20:1	22:1	22:1 n-11	22:1 n-9	24:1	Monounsatd
Cows' milk																
84	**Skimmed milk**, pasteurised, average	Tr	0	0	0	Tr	0	0.06	0.06	Tr	Tr	0	0	0	0	0
85	**Semi-skimmed milk**, pasteurised, average	Tr	0	0.02	0	0.02	0	0.34	0.34	Tr	Tr	0	0	0	0	0.05
86	**Whole milk**, pasteurised, average	0.01	0	0.04	0	0.07	0.01	0.80	0.79	0.01	0.01	0	0	0	0	0.11
87	sterilised	0.01	0	0.03	0	0.06	0.01	0.79	0.73	0.11	0.01	0	0	0	0	0.11
88	UHT	0.01	0	0.03	0	0.06	0.01	0.80	0.75	0.12	0.01	0	0	0	0	0.12
89	**Breakfast milk**, pasteurised, average	0.01	0	0.03	0	0.07	0.01	0.86	0.80	0.12	0.01	0	0	0	0	0.13
90	**Organic semi-skimmed milk**, pasteurised	0.01	0	0.02	0	0.03	Tr	0.33	0.31	0.04	Tr	0	0	0	0	0.04
91	**Evaporated milk**, light	0.01	0	0.04	0	0.06	0.01	0.82	0.78	0.12	0.01	0	0	0	0	0.13
Other milks, dairy products and substitutes																
92	**Coffee compliment** powder	0	0	0	0	0	0	0.18	0.16	0	0	0.02	0	0.02	0	0.08
93	low fat	0	0	0	0	0	0	0.24	0.36	0.07	0	0	0	0	0	0.22
94	**Coffee whitener** liquid, with skimmed milk and non milk fat	0	0	0	0	0	0	0.13	0.20	0.03	0.03	0	0	0	0	0.12
95	with glucose syrup and vegetable fat	0	0	0	0	0	0	0.25	0.50	0.03	0.03	0	0	0	0.01	0.29
96	**Chocolate flavoured milk**, pasteurised	0	0	0.01	0	0.01	0	0.13	0.18	0.01	0.01	0	0	0	0	0.07
97	**Flavoured milk**, pasteurised	0	0	0.01	0	0.02	0	0.26	0.24	0.03	0	0	0	0	0	0.03
98	**Goats' milk**	0	0	0	0	0.05	0.01	0.76	0.75	0	0	0.01	0	0.01	0	0.11

Polyunsaturated fatty acids, g per 100g food

No.	Food	cis n-6					cis n-3					trans
		18:2	18:3	20:3	20:4	22:4	18:3	18:4	20:5	22:5	22:6	Polyunsatd
Cows' milk												
84	**Skimmed milk**, pasteurised, average	0.01	0	0	0	0	Tr	0	0	Tr	0	Tr
85	**Semi-skimmed milk**, pasteurised, average	0.03	0	0	0	0	0.01	0	0	Tr	0	0.02
86	**Whole milk**, pasteurised, average	0.07	0	0	0	0	0.02	0	0	Tr	0	0.03
87	sterilised	0.07	0	0	0	0	0.02	0	0	Tr	0	0.03
88	UHT	0.06	0	0	0	0	0.02	0	0	Tr	0	0.03
89	**Breakfast milk**, pasteurised, average	0.08	0	0	0.01	0	0.02	0	0	Tr	0	0.03
90	**Organic semi-skimmed milk**, pasteurised	0.04	0	0	0	0	0.01	0	0	Tr	0	0.01
91	**Evaporated milk**, light	0.06	0	0	0.01	0	0.02	0	0	Tr	0	0.04
Other milks, dairy products and substitutes												
92	**Coffee compliment** powder	0.02	0	0	0	0	0.02	0	0	0	0	0
93	low fat	0.04	0	0	0	0	0	0	0	0	0	0
94	**Coffee whitener** liquid, with skimmed milk and non milk fat	0.02	0	0	0	0	0	0	0	0	0	0.01
95	with glucose syrup and vegetable fat	0.02	0	0	0	0	0	0	0	0	0	0.01
96	**Chocolate flavoured milk**, pasteurised	0.02	0	0	0	0	0	0	0	Tr	0	0
97	**Flavoured milk**, pasteurised	0.02	0	0	0	0	0.01	0	0	Tr	0	0.01
98	**Goats milk**	0.10	0	0	0	0	0.03	0	0	0	0	0.03

Fat and total fatty acids, g per 100g food

No.	Food	Description	Total fat	Satd	cis-Mono unsatd	Polyunsatd Total cis	n-6	n-3	Total trans	Total branched
Other milks, dairy products and substitutes continued										
99	**Human milk**, mature	Data from Institute of Human Nutrition and Brain Chemistry	4.1	1.92	1.47	0.49	0.41	0.08	0	0
100	**Lassi**, sweetened	5 samples including takeaway	0.9	0.59	0.19	0.02	0.02	0.01	0.02	0
101	**Milkshake**, thick, takeaway	10 samples, 3 brands including chocolate, vanilla, and banana	1.8	1.17	0.37	0.05	0.04	0.01	0.06	0
102	**Sheep's milk**	3 samples, 2 brands	5.8	3.58	1.22	0.21	0.29	0.06	0.38	0.01
103	**Soya, non-dairy alternative to milk**	10 samples, 8 brands	1.6	0.24	0.33	0.93	0.81	0.12	Tr	0
104	**Tea whitener** powder	3 samples of the same brand	14.8	13.73	0.21	0.04	0.04	0	0.12	0
Creams										
105	**Single cream**, fresh, pasteurised	Fatty acids calculated from whole cows' milk	19.1	12.15	4.54	0.47	0.49	0.11	0.68	0
106	**Double cream**, fresh, pasteurised	Average of 22 samples of summer and winter cream	53.7	33.39	12.33	1.49	1.34	0.48	1.83	0
107	**Creme fraiche**	9 samples, 5 brands	40.0	27.10	7.96	0.88	0.83	0.29	0.77	0
108	**Dairy cream**, UHT, canned spray	10 samples, 6 brands	24.2	15.20	5.43	0.63	0.56	0.15	0.81	0
Imitation creams										
109	**Elmlea**, single [a]	5 samples	14.5	9.19	2.85	1.22	0.88	0.36	0.37	0
110	whipping [a]	4 samples	29.9	24.64	2.13	0.85	N	N	N	0
111	double [b]	4 samples	35.7	24.34	5.71	2.70	1.98	0.79	0.89	0

a Contains 0.02g unidentified fatty acids per 100g food
b Contains 0.03g unidentified fatty acids per 100g food

No.	Food	4:0	6:0	8:0	10:0	12:0	14:0	15:0	16:0	17:0	18:0	20:0	22:0	24:0
Other milks, dairy products and substitutes continued														
99	**Human milk**, mature	0	0	0	0.06	0.09	0.23	0	1.12	0	0.38	0.02	0.01	0.01
100	**Lassi**, sweetened	0.04	0.02	0.01	0.03	0.03	0.10	Tr	0.26	0.01	0.08	0	Tr	0
101	**Milkshake**, thick, takeaway	0.07	0.04	0.02	0.05	0.06	0.19	0.02	0.52	0.01	0.18	0	0	0
102	**Sheep's milk**	0.17	0.15	0.15	0.41	0.22	0.53	0.09	1.19	0.04	0.61	0.02	0.01	0.01
103	**Soya, non-dairy alternative to milk**	0	0	0	0	0	Tr	0	0.16	0	0.06	0.01	0.01	0
104	**Tea whitener** powder	0	0.06	0.71	0.52	5.54	1.77	0	1.36	0.01	3.71	0.04	0.01	0.01
Creams														
105	**Single cream**, fresh, pasteurised	*0.69*	*0.43*	*0.24*	*0.54*	*0.73*	*2.02*	*0.20*	*5.21*	*0.10*	*2.00*	*Tr*	*0*	*0*
106	**Double cream**, fresh, pasteurised a	2.16	1.22	0.68	1.44	1.68	5.28	0.56	14.02	0.38	5.77	0.09	0.04	0
107	**Creme fraiche** b	1.68	1.02	0.59	1.32	1.54	4.56	0.43	12.46	0.23	3.12	0.03	0	0
108	**Dairy cream**, UHT, canned spray c	0.99	0.56	0.32	0.66	0.82	2.45	0.24	6.59	0.16	2.35	0.03	0	0
Imitation creams														
109	**Elmlea**, single	0.02	0.04	0.37	0.32	4.14	1.44	0.01	1.32	0.01	1.49	0.03	0	0
110	whipping	0.02	0.15	1.55	1.21	11.41	4.41	N	2.96	N	2.89	0.04	0	0
111	double	0.02	0.10	1.04	0.88	11.18	3.93	N	3.06	0.02	4.06	0.06	0	0

a Contains 0.03g 11:0, 0.04g 13:0 per 100g food
b Contains 0.03g 11:0, 0.05g 13:0 per 100g food
c Contains 0.01g 11:0, 0.02g 13:0 per 100g food

Milk and milk products

Monounsaturated fatty acids, g per 100g food

No.	Food	cis							cis/trans		cis		cis/trans		cis	trans
		10:1	12:1	14:1	15:1	16:1	17:1	18:1	18:1 n-9	18:1 n-7	20:1	22:1	22:1 n-11	22:1 n-9	24:1	Monounsatd
	Other milks, dairy products and substitutes continued															
99	**Human milk**, mature	0	0	0	0	0.12	0	1.32	1.32	0	0.02	0	0	0	0.01	0
100	**Lassi**, sweetened	0	0	0.01	0	0.01	Tr	0.16	0.15	0.02	0	0	0	0	0	0.02
101	**Milkshake**, thick, takeaway	0	0	0.02	0	0.03	0.01	0.31	0.30	0.04	0	0	0	0	0	0.04
102	**Sheep's milk**	0	0	0	0	0.06	0.02	1.12	1.04	0.08	0.01	0.01	0	0	0	0.25
103	**Soya, non-dairy alternative to milk**	0	0	0	0	0	0	0.33	0.31	0.02	0	0	0	0	0	Tr
104	**Tea whitener** powder	0	0	0	0	0	0	0.21	0.28	0.03	0	0	0	0	0	0.12
	Creams															
105	**Single cream**, fresh, pasteurised	0.04	N	0.20	0	0.32	0.05	3.90	3.63	0.58	0.04	0	0	0	N	0.53
106	**Double cream**, fresh, pasteurised	0.14	0.04	0.44	0	0.81	0.12	10.61	9.87	1.50	0.13	0	0	0	0.04	1.48
107	**Creme fraiche**	0.13	0.03	0.39	0	0.68	0.11	6.48	6.09	0.62	0.04	0	0	0	0.10	0.59
108	**Dairy cream**, UHT, canned spray	0.07	0.02	0.22	0	0.38	0.07	4.64	4.39	0.56	0	0	0	0	0.03	0.64
	Imitation creams															
109	**Elmlea**, single	0	0	0	0	0	0	2.83	2.87	0.24	0	0	0	0	0.02	0.34
110	whipping	0	0	0	0	0	0	2.07	2.45	0.23	0.03	0	0	0	0.03	0.68
111	double	0	0	0	0	0	0	5.68	5.89	0.52	0	0	0	0	0.02	0.81

Milk and milk products

Polyunsaturated fatty acids, g per 100g food

| No. | Food | cis n-6 | | | | | cis n-3 | | | | | trans |
		18:2	18:3	20:3	20:4	22:4	18:3	18:4	20:5	22:5	22:6	Polyunsatd
Other milks, dairy products and substitutes continued												
99	**Human milk**, mature a	0.36	0	0.01	0.02	Tr	0.03	0	0.02	0.01	0.02	0
100	**Lassi**, sweetened	0.01	0	0	0	0	0	0	0	Tr	0	0.01
101	**Milkshake**, thick, takeaway	0.03	0	0	0	0	0.01	0	0	Tr	0	0.01
102	**Sheep's milk**	0.15	0	0	0	0	0.06	0	0	0	0	0.13
103	**Soya, non-dairy alternative to milk**	0.81	0	0	0	0	0.12	0	0	0	0	Tr
104	**Tea whitener** powder	0.04	0	0	0	0	0	0	0	0	0	0
Creams												
105	**Single cream**, fresh, pasteurised	0.37	0	0	0	0	0.10	0	0	0	0	0.16
106	**Double cream**, fresh, pasteurised b	0.94	0	0.03	0.07	0	0.31	0	0.04	0.05	0	0.36
107	**Creme fraiche**	0.62	0	0.02	0.03	0	0.15	0.05	0	0	0	0.19
108	**Dairy cream**, UHT, canned spray	0.39	0.01	0.02	0.02	0	0.12	0	0.02	0.02	0	0.17
Imitation creams												
109	**Elmlea**, single	0.87	0.01	0.01	0	0	0.34	0	0.01	0	0	0.03
110	whipping	0.68	0	0	0	0	0.15	0	0.01	0	0	N
111	double	1.98	0	0	0	0	0.75	0	0	0	0	0.07

a Contains 0.01g 20:2 per 100g food
b Contains 0.04g 21:5 per 100g food

45

Milk and Milk products

Fat and total fatty acids, g per 100g food

No.	Food	Description	Total fat	Satd	cis-Mono unsatd	Total cis	Polyunsatd n-6	n-3	Total trans	Total branched
Imitation creams continued										
112	**Non-dairy cream**, UHT, canned spray	5 samples, 4 brands	22.2	19.02	0.80	0.16	0.16	0.03	1.17	0
113	**Simply Double dessert topping**	4 samples	28.2	24.80	0.94	0.03	0.17	0	1.04	0
114	**Soya dessert topping** a	6 samples, 2 brands	18.7	2.76	4.02	10.68	9.59	1.30	0.22	0
115	**Tip Top dessert topping**	4 samples	6.5	5.94	0.09	0.10	0.08	0.02	0.07	0

a Contains 0.01g unidentified fatty acids per 100g food

Milk and milk products

Saturated fatty acids, g per 100g food

No.	Food	4:0	6:0	8:0	10:0	12:0	14:0	15:0	16:0	17:0	18:0	20:0	22:0	24:0
	Imitation creams continued													
112	**Non-dairy cream**, UHT, canned spray	0	0.08	0.89	0.76	8.94	3.30	0	2.05	0	2.92	0.04	0	0.01
113	**Simply Double dessert topping**	0	0.10	1.22	1.01	11.41	4.32	0	2.62	0	4.05	0.04	0	0
114	**Soya dessert topping**	0	0	0	0	0.01	0.02	0	1.97	0.02	0.64	0.05	0.04	0.02
115	**Tip Top dessert topping**	0	0.05	0.51	0.37	2.56	1.03	0	0.63	0	0.78	0.01	0	0

Monounsaturated fatty acids, g per 100g food

No.	Food		cis							cis/trans		cis		cis/trans		cis	trans
		10:1	12:1	14:1	15:1	16:1	17:1	18:1	18:1 n-9	18:1 n-7	20:1	22:1	22:1 n-11	22:1 n-9	24:1	Monounsatd
Imitation creams continued																	
112	**Non-dairy cream**, UHT, canned spray	0	0	0	0	0	0	0.78	1.80	0.12	0.01	0.01	0.01	0	0	1.17
113	**Simply Double dessert topping**	0	0	0	0	0	0	0.90	1.40	0.25	0	0	0	0	0.04	0.99
114	**Soya dessert topping**	0	0	0	0	0.02	0.01	3.98	3.74	0.24	0	0	0	0	0	0.01
115	**Tip Top dessert topping**	0	0	0	0	0	0	0.09	0.12	0.02	0	0	0	0	0	0.07

Polyunsaturated fatty acids, g per 100g food

No.	Food	cis n-6					cis n-3					trans
		18:2	18:3	20:3	20:4	22:4	18:3	18:4	20:5	22:5	22:6	Polyunsatd
	Imitation creams continued											
112	**Non-dairy cream**, UHT, canned spray	0.10	0	0	0	0	0.02	0	0	0	0	0
113	**Simply Double dessert topping**	0.01	0.02	0	0	0	0	0	0	0	0	0.06
114	**Soya dessert topping**	9.55	0	0	0	0	1.17	0	0	0	0	0.21
115	**Tip Top dessert topping**	0.08	0	0	0	0	0.01	0	0	0	0	0

No.	Food	Description	Total fat	Satd	cis-Mono unsatd	Polyunsatd Total cis	n-6	n-3	Total trans	Total branched
	Infant formulas									
	Whey-based modified milks									
116	**Milupa Aptamil**, reconstituted	Manufacturer's data (Milupa)	3.6	1.74	1.29	0.50	0.44	0.06	Tr	0
117	**Cow & Gate Premium**, reconstituted	Manufacturer's data (Cow & Gate)	3.5	1.30	1.53	0.46	0.39	0.07	Tr	0
118	**Farley's First Milk**, reconstituted	Manufacturer's data (Farley's)	3.8	1.31	1.88	0.45	0.40	0.06	Tr	0
119	**SMA Gold**, reconstituted	Manufacturer's data (Wyeth)	3.6	1.62	1.34	0.64	0.60	0.03	Tr	0
	Non-whey-based modified milks									
120	**Cow & Gate Plus**, reconstituted	Manufacturer's data (Cow & Gate)	3.4	1.25	1.53	0.45	0.38	0.07	Tr	0
121	**Farley's Second Milk**, reconstituted	Manufacturer's data (Farley's)	2.9	1.02	1.48	0.27	0.24	0.03	Tr	0
122	**Milupa Milumil**, reconstituted	Manufacturer's data (Milupa)	3.1	1.30	1.20	0.58	0.53	0.05	Tr	0
123	**SMA White**, reconstituted	Manufacturer's data (Wyeth)	3.6	1.62	1.34	0.64	0.60	0.03	Tr	0
	Soya-based modified milks									
124	**Cow & Gate Infasoy**, reconstituted	Manufacturer's data (Cow & Gate)	3.6	1.41	1.67	0.50	0.42	0.08	Tr	0
125	**Farley's Soya Formula**, reconstituted	Manufacturer's data (Farley's)	3.8	1.33	1.94	0.36	0.32	0.04	Tr	0
126	**Prosobee**, reconstituted	Manufacturer's data (Mead Johnson Nutritionals)	3.6	1.46	1.30	0.65	0.59	0.06	Tr	0
127	**SMA Wysoy**, reconstituted	Manufacturer's data (Wyeth)	3.6	1.62	1.34	0.64	0.60	0.03	Tr	0

Milk and milk products

Saturated fatty acids, g per 100ml feed

No.	Food	4:0	6:0	8:0	10:0	12:0	14:0	15:0	16:0	17:0	18:0	20:0	22:0	24:0
	Infant formulas													
	Whey-based modified milks													
116	**Milupa Aptamil**, reconstituted	0.04	0.02	0.05	0.06	0.22	0.21	0	0.91	0	0.22	0.01	Tr	Tr
117	**Cow & Gate Premium**, reconstituted	0	0.01	0.05	0.04	0.32	0.14	0	0.59	0	0.01	0.01	0.01	0
118	**Farley's First Milk**, reconstituted [a]	Tr	Tr	0.02	0.03	0.42	0.15	0	0.51	0	0.14	0.03	0.01	Tr
119	**SMA Gold**, reconstituted	0	0	0.06	0.06	0.36	0.16	0	0.79	0	0.16	0.01	0.01	0
	Non-whey-based modified milks													
120	**Cow & Gate Plus**, reconstituted	0	0.01	0.05	0.04	0.31	0.14	0	0.57	0	0.11	0.01	0.01	0
121	**Farley's Second Milk**, reconstituted [b]	Tr	Tr	0.02	0.03	0.34	0.12	0	0.39	0	0.10	0.02	0.01	Tr
122	**Milupa Milumil**, reconstituted	0	Tr	0.02	0.02	0.14	0.08	0	0.91	0	0.11	0.01	0.01	Tr
123	**SMA White**, reconstituted	0	0	0.06	0.06	0.36	0.16	0	0.79	0	0.16	0.01	0.01	0
	Soya-based modified milks													
124	**Cow & Gate Infasoy**, reconstituted	0	Tr	0.05	0.04	0.33	0.15	0	0.68	0	0.13	0.01	0.01	0
125	**Farley's Soya Formula**, reconstituted [a]	0	Tr	0.03	0.03	0.44	0.16	0	0.51	0	0.13	0.02	0.01	Tr
126	**Prosobee**, reconstituted	0	Tr	0.05	0.04	0.32	0.14	0	0.75	0	0.15	0.01	0	0
127	**SMA Wysoy**, reconstituted	0	0	0.06	0.06	0.36	0.16	0	0.79	0	0.16	0.01	0.01	0

[a] Contains 0.03g other saturated fatty acids per 100ml feed
[b] Contains 0.02g other saturated fatty acids per 100ml feed

Milk and milk products

Monounsaturated fatty acids, g per 100ml feed

No. Food	cis 10:1	12:1	14:1	15:1	16:1	17:1	18:1	cis/trans 18:1 n-9	18:1 n-7	cis 20:1	22:1	cis/trans 22:1 n-11	22:1 n-9	cis 24:1	trans Monounsatd
Infant formulas															
Whey-based modified milks															
116 **Milupa Aptamil**, reconstituted	0	0	0	0	0.05	0	1.23	N	N	0.02	0	0	0	0	Tr
117 **Cow & Gate Premium**, reconstituted	0	0	0	0	0.01	0	1.50	N	N	0.02	Tr	0	Tr	Tr	Tr
118 **Farley's First Milk**, reconstituted [a]	0	0	0	0	Tr	0	1.85	N	N	0.02	Tr	0	0	0	Tr
119 **SMA Gold**, reconstituted	0	0	0	0	0	0	1.33	N	N	0.01	0	0	0	0	Tr
Non-whey-based modified milks															
120 **Cow & Gate Plus**, reconstituted	0	0	0	0	0.01	0	1.50	N	N	0.02	Tr	0	0	Tr	Tr
121 **Farley's Second Milk**, reconstituted [b]	0	0	0	0	Tr	0	1.47	N	N	0.01	Tr	0	0	0	Tr
122 **Milupa Milumil**, reconstituted	0	0	0	0	Tr	0	1.19	N	N	0.01	0	0	0	0	Tr
123 **SMA White**, reconstituted	0	0	0	0	Tr	0	1.33	N	N	0.01	0	0	0	0	Tr
Soya-based modified milks															
124 **Cow & Gate Infasoy**, reconstituted	0	0	0	0	0.01	0	1.64	N	N	0.02	Tr	0	0	Tr	Tr
125 **Farley's Soya Formula**, reconstituted [b]	0	0	0	0	Tr	0	1.93	N	N	0.01	Tr	0	0	0	Tr
126 **Prosobee**, reconstituted	0	0	0	0	Tr	0	1.29	N	N	0	0	0	0	0	Tr
127 **SMA Wysoy**, reconstituted	0	0	0	0	Tr	0	1.33	N	N	0.01	0	0	0	0	Tr

a Contains 0.02g other monounsaturated fatty acids per 100ml feed
b Contains 0.01g other monounsaturated fatty acids per 100ml feed

Milk and milk products

Polyunsaturated fatty acids, g per 100ml feed

No. Food	18:2	18:3	*cis* n-6			18:3	*cis* n-3				*trans*
			20:3	20:4	22:4		18:4	20:5	22:5	22:6	Polyunsatd
Infant formulas											
Whey-based modified milks											
116 **Milupa Aptamil**, reconstituted	0.42	0.01	Tr	0.01	Tr	0.05	0	0	Tr	0.01	Tr
117 **Cow & Gate Premium**, reconstituted	0.38	0.07	0	0	0	0.07	0	0	0	0	Tr
118 **Farley's First Milk**, reconstituted	0.35	0.03	0	Tr	0	0.04	Tr	Tr	Tr	0.02	Tr
119 **SMA Gold**, reconstituted	0.58	0.02	0	0	0	0.03	0	0	0	0	Tr
Non-whey-based modified milks											
120 **Cow & Gate Plus**, reconstituted	0.37	0.01	N	N	0	0.07	0	N	N	N	Tr
121 **Farley's Second Milk**, reconstituted	0.24	0	0	0	0	0.03	0	0	0	0	Tr
122 **Milupa Milumil**, reconstituted	0.51	0.01	0	0	0	0.05	0	0	0	0	Tr
123 **SMA White**, reconstituted	0.58	0.02	0	0	0	0.03	0	0	0	0	Tr
Soya-based modified milks											
124 **Cow & Gate Infasoy**, reconstituted	0.42	Tr	0	0	0	0.08	0	0	N	N	Tr
125 **Farley's Soya Formula**, reconstituted	0.32	0	0	0	0	0.04	0	0	0	0	Tr
126 **Prosobee**, reconstituted	*0.59*	0	0	0	0	*0.06*	0	0	0	0	Tr
127 **SMA Wysoy**, reconstituted	0.58	0.02	0	0	0	0.03	0	0	0	0	Tr

Fat and total fatty acids, g per 100g food

No.	Food	Description	Total fat	Satd	cis-Mono unsatd	Polyunsatd Total cis	n-6	n-3	Total trans	Total branched
Cheese										
128	**Brie**	Data from Institute of Human Nutrition and Brain Chemistry	26.9	16.87	5.96	0.37	0.29	0.24	1.57	0.19
129	**Cheddar**, average	Mixed samples of English, Irish and New Zealand cheese	32.7	19.25	7.14	0.77	0.99	0.28	2.10	0.74
130	**Cottage cheese** [a]	Fatty acids calculated from whole cows milk	3.9	2.48	0.93	0.09	0.09	0.03	0.14	0.07
131	**Cream cheese** [b]	Data from Institute of Human Nutrition and Brain Chemistry	47.5	29.73	11.08	0.65	0.66	0.27	2.76	0.34
132	**Edam** [c]	Data from Institute of Human Nutrition and Brain Chemistry	24.5	15.48	5.21	0.37	0.37	0.12	1.06	0.17
133	**Processed cheese slices**	11 samples, 6 brands	23.0	14.32	5.56	0.48	0.63	0.20	1.13	0.04
134	reduced fat	10 samples, 7 brands	13.3	8.10	3.24	0.37	0.34	0.11	0.43	0.28
135	**Stilton, blue** [d]	Data from Institute of Human Nutrition and Brain Chemistry	34.0	20.55	6.76	0.68	0.81	0.13	1.30	0.22
Yogurts										
136	**Whole milk yogurt**, plain [e]	Fatty acids calculated from whole cows milk	*3.0*	*1.91*	*0.71*	*0.07*	*0.05*	*0.02*	*0.11*	*0.05*
137	fruit	10 samples, 6 brands including thick and creamy types	3.0	2.01	0.67	0.09	0.07	0.02	0.07	0.02
138	–, infant	8 samples, 3 brands, assorted flavours	3.7	2.45	0.85	0.09	0.08	0.02	0.08	0.02
139	**Low fat yogurt**, plain	8 samples, 5 brands	1.0	0.66	0.23	0.03	0.03	Tr	0.02	0.01
140	hazelnut	8 samples, 5 brands	1.5	0.57	0.75	0.09	0.08	0.01	N	0.01
141	toffee	8 samples, 4 brands	0.9	0.61	0.20	0.02	0.02	0.01	0.01	0.01

[a] Contains 0.13g unidentified fatty acids per 100g food
[b] Contains 0.71g unidentified fatty acids per 100g food
[c] Contains 0.54g unidentified fatty acids per 100g food
[d] Contains 1.02g unidentified fatty acids per 100g food
[e] Contains 0.10g unidentified fatty acids per 100g food

Milk and milk products

Saturated fatty acids, g per 100g food

No.	Food	4:0	6:0	8:0	10:0	12:0	14:0	15:0	16:0	17:0	18:0	20:0	22:0	24:0
Cheese														
128	**Brie**	0.82	0.52	0.31	0.72	0.90	2.87	0.66	6.94	N	3.13	N	N	0
129	**Cheddar**, average	0.53	0.46	0.37	0.96	1.21	3.49	0.37	7.63	0.25	3.89	0.06	0.03	0
130	**Cottage cheese**	0.14	0.09	0.05	0.11	0.15	0.41	0.04	1.06	0.02	0.41	N	N	0
131	**Cream cheese**	1.45	0.91	0.54	1.27	1.59	5.06	1.17	12.22	N	5.52	N	N	0
132	**Edam**	0.39	0.24	0.37	0.69	0.94	2.73	0.59	6.93	0.11	2.49	0.02	N	0
133	**Processed cheese slices**	0.70	0.46	0.28	0.57	0.72	2.24	0.48	6.04	0.13	2.61	0.04	0.04	0.02
134	reduced fat	0.33	0.23	0.15	0.34	0.42	1.37	0.15	3.68	0.09	1.32	0.02	0.01	0
135	**Stilton, blue**	N	N	0.57	1.01	1.28	3.72	0.77	9.64	0.32	3.18	0.05	0.01	0
Yogurts														
136	**Whole milk yogurt**, plain	0.11	0.07	0.04	0.08	0.12	0.32	0.03	0.82	0.02	0.31	0	0	0
137	fruit	0.10	0.06	0.03	0.09	0.10	0.32	0.03	0.93	0.02	0.30	0	0	0
138	–, infant	0.12	0.07	0.04	0.09	0.13	0.40	0.04	1.13	0.03	0.36	0.03	0	0
139	**Low fat yogurt**, plain	0.04	0.02	0.01	0.02	0.03	0.10	0.01	0.30	Tr	0.11	0.01	0	0
140	hazelnut	0.03	0.02	0.01	0.02	0.03	0.08	0.01	0.27	Tr	0.09	0.01	Tr	0
141	toffee	0.03	0.01	0.01	0.02	0.03	0.10	0.01	0.29	0.01	0.09	0.01	0	0

Milk and milk products

Monounsaturated fatty acids, g per 100g food

No.	Food	cis							cis/trans			cis	cis/trans		cis	trans
		10:1	12:1	14:1	15:1	16:1	17:1	18:1	18:1 n-9	18:1 n-7	20:1	22:1	22:1 n-11	22:1 n-9	24:1	Monounsatd
Cheese																
128	**Brie** [a]	0	0	0.24	0	0.41	0	5.31	5.28	0.04	0	0	0	0	0	1.41
129	**Cheddar**, average	0.09	0	0.25	0	0.43	0.09	6.03	6.27	1.11	0.19	0.06	0	0.06	0	1.58
130	**Cottage cheese** [b]	0.01	0	0.04	0	0.07	0.01	0.80	0.74	0.12	0.01	0	0	0	0	0.11
131	**Cream cheese** [c]	0	0	0.42	0	0.72	N	9.36	9.30	0.07	0	0	0	0	0	2.49
132	**Edam**	0	0	0.11	0	0.37	N	4.56	4.53	0.03	0.01	0	0	0	0	0.92
133	**Processed cheese slices**	0	0	0	0	0.41	0.07	5.00	4.72	0.28	0.02	0.02	0	0.04	0	0.78
134	reduced fat	0.03	0.01	0.11	0	0.37	0.04	2.65	2.48	0.35	0.03	0	0	0	0	0.35
135	**Stilton, blue**	0	0	0	0	0.44	0	6.26	6.22	0.04	0.06	0	0	0	0	1.04
Yogurts																
136	**Whole milk yogurt**, plain	0.01	0	0.03	0	0.05	0.01	0.61	0.57	0.09	0.01	0	0	0	0	0.08
137	fruit	0.02	0	0.05	0.01	0.04	0.01	0.57	0.57	0	0	0	0	0	0	0.06
138	-, infant	0.02	0	0.06	0	0.07	Tr	0.69	0.69	0	0.01	0	0	0	0	0.07
139	**Low fat yogurt**, plain	0	0	0.01	0	0.01	Tr	0.19	0.19	0	0	0	0	0	0	0.02
140	hazelnut	0	0	0.01	0	0.02	Tr	0.72	0.72	0	0	0	0	0	0	N
141	toffee	0	0	0.01	0	0.02	Tr	0.17	0.17	0	0	0	0	0	0	0.01

[a] Contains 0.32g other monounsaturated fatty acids per 100g food
[b] Contains 0.57g other monounsaturated fatty acids per 100g food
[c] Contains 0.15g other monounsaturated fatty acids per 100g food

Milk and milk products

Polyunsaturated fatty acids, g per 100g food

No.	Food	cis n-6					cis n-3					trans
		18:2	18:3	20:3	20:4	22:4	18:3	18:4	20:5	22:5	22:6	Polyunsatd
Cheese												
128	**Brie**	0.22	0	0	0	0	0.15	0	0	0	0	0.16
129	**Cheddar**, average a	0.31	0.18	0	0.03	0	0.23	0	0	0	0	0.53
130	**Cottage cheese**	0.07	0	0	0	0	0.02	0	0	0	0	0.03
131	**Cream cheese**	0.39	0	0	0	0	0.27	0	0	0	0	0.28
132	**Edam**	0.24	0	0.01	0	0	0.12	0	0	0	0	0.13
133	**Processed cheese slices**	0.28	0	0	0	0	0.20	0	0	0	0	0.35
134	reduced fat b	0.22	0.03	0.01	0.02	0	0.07	0	0.01	0.01	0	0.08
135	**Stilton, blue**	0.52	0	0.02	0.02	0	0.13	0	0	0	0	0.26
Yogurts												
136	**Whole milk yogurt**, plain	0.05	0	0	0	0	0.02	0	0	0	0	0.03
137	fruit	0.06	0	0	0	0	0.02	0	0	0	0	0.01
138	-, infant	0.07	0	0	0	0	0.02	0	0	0	0	0.01
139	**Low fat yogurt**, plain	0.03	0	0	0	0	Tr	0	0	0	0	Tr
140	hazelnut	0.08	0	0	0	0	0.01	0	0	0	0	Tr
141	toffee	0.02	0	0	0	0	0.01	0	0	0	0	Tr

a Contains 0.03g 16:2 per 100g food
b Contains 0.01g 22:2 per 100g food

Milk and milk products

Fat and total fatty acids, g per 100g food

No.	Food	Description	Total fat	Satd	cis-Mono unsatd	Polyunsatd Total cis	n-6	n-3	Total trans	Total branched
Yogurts continued										
142	**Reduced fat yogurt**, frozen	2 samples, different brands	2.0	1.27	0.47	0.05	0.05	0.01	0.06	0.03
143	**Virtually fat free/diet yogurt**, fruit, twin pot	5 samples, 2 brands	0.1	0.05	0.02	0.02	0.01	0.01	0	0
144	**Fromage frais**, plain	5 samples, 3 brands	8.0	5.53	1.69	0.17	0.15	0.02	0.11	0
145	fruit, children's	13 samples, 6 brands including strawberry, raspberry, peach, and apricot	4.8	3.26	1.08	0.07	0.08	0.01	0.09	0
146	**Greek style yogurt**, plain	7 samples, 6 brands; cow's milk	10.2	6.75	2.35	0.25	0.23	0.05	0.21	0.06
147	**Soya, alternative to yogurt**, fruit	3 samples of the same brand; strawberry	1.8	0.28	0.36	1.06	0.92	0.13	0	0
Ice creams										
148	**Ice cream**, dairy	5 samples	8.6	5.07	1.79	0.20	0.23	0.01	0.38	0.27
149	reduced calorie	5 samples, 2 brands	5.2	3.22	1.26	0.12	0.18	0	0.26	0
150	**Ice cream bar**, chocolate coated	10 samples, different brands including Mars, Bounty, Snickers	23.3	12.48	7.30	1.11	1.05	0.07	0.68	0.10
Puddings and chilled desserts										
151	**Cheesecake**, fruit, individual	8 samples, 3 brands including strawberry, apricot, blackcurrant, and cherry	12.3	7.54	3.28	0.49	0.42	0.09	0.23	0.09
152	**Chocolate dairy dessert**, individual	8 samples, 4 brands including milk chocolate, and caramel, and white chocolate dessert pots	10.7	6.26	2.86	0.32	0.30	0.06	0.46	0.13
153	**Creme caramel**	8 samples, 3 brands	1.6	0.89	0.34	0.02	0.04	0.01	0.16	0.04
154	**Custard**, ready to eat	5 samples, 3 brands; canned and tetra pack	2.9	1.89	0.69	0.06	0.05	0.01	0.07	0

Milk and milk products

Saturated fatty acids, g per 100g food

No.	Food	4:0	6:0	8:0	10:0	12:0	14:0	15:0	16:0	17:0	18:0	20:0	22:0	24:0
Yoghurts continued														
142	**Reduced fat yogurt**, frozen	0.05	0.03	0.02	0.05	0.06	0.18	0.02	0.53	0.01	0.31	Tr	0	0
143	**Virtually fat free/diet yogurt**, fruit, twin pot	0	0	0	0	0	0	0	0.02	0	0.01	0	0	0
144	**Fromage frais**, plain	0.27	0.15	0.09	0.21	0.27	0.90	0.09	2.75	0.05	0.69	0.05	0	0
145	fruit, children's	0.17	0.10	0.06	0.12	0.17	0.53	0.06	1.55	0.02	0.43	0.04	0	0
146	**Greek style yogurt**, plain	0.33	0.20	0.11	0.26	0.35	1.11	0.11	3.12	0.07	1.00	0.08	0	0
147	**Soya, alternative to yogurt**, fruit	0	0	0	0	0	0.01	0	0.18	0	0.08	0.01	0.01	0
Ice creams														
148	**Ice cream**, dairy	0.28	0.18	0.14	0.27	0.30	0.86	0.10	2.19	0.06	0.67	0.01	0	0
149	reduced calorie	0.14	0.08	0.05	0.11	0.17	0.46	0.09	1.29	0.03	0.77	0.01	0.01	0
150	**Ice cream bar**, chocolate coated	0.12	0.07	0.06	0.14	0.22	0.81	0.10	5.71	0.10	5.14	0	0	0
Puddings and chilled desserts														
151	**Cheesecake**, fruit, individual	0.34	0.17	0.10	0.22	0.33	1.06	0.11	3.93	0.05	1.11	0.10	0.03	0
152	**Chocolate dairy dessert**, individual	0.17	0.11	0.07	0.16	0.21	0.65	0.07	2.75	0.05	1.96	0.05	0	0
153	**Creme caramel**	0.02	0.01	0.01	0.03	0.05	0.15	0.02	0.42	0.01	0.17	0	0	0
154	**Custard**, ready to eat	0.08	0.05	0.03	0.07	0.10	0.32	0.04	0.88	0.01	0.30	0.02	0	0

Milk and milk products

Monounsaturated fatty acids, g per 100g food

No.	Food	10:1	12:1	14:1	cis 15:1	16:1	17:1	18:1	cis/trans 18:1 n-9	18:1 n-7	cis 20:1	cis 22:1	cis/trans 22:1 n-11	22:1 n-9	cis 24:1	trans Monounsatd
Yogurts continued																
142	**Reduced fat yogurt**, frozen	0	0	0.01	0	0.03	0.01	0.41	0.35	0.05	0	0	0	0	0	0.05
143	**Virtually fat free/diet yogurt**, fruit, twin pot	0	0	0	0	0	0	0.02	0.02	0	0	0	0	0	0	0
144	**Fromage frais**, plain	0.07	0	0.13	0.02	0.13	0.03	1.31	1.31	0	0	0	0	0	0	0.11
145	fruit, children's	0.01	0	0.08	0.02	0.09	N	0.88	0.88	0	0	0	0	0	0	0.07
146	**Greek style yogurt**, plain	0.06	0	0.15	0	0.19	0	1.91	1.91	0	0.04	0	0	0	0	0.18
147	**Soya, alternative to yogurt**, fruit	0	0	0	0	0	0	0.35	0.35	0	0	0	0	0	0	0
Ice creams																
148	**Ice cream**, dairy	0.03	0.01	0.03	0	0.11	0.02	1.53	0.83	0.70	0.01	0	0	0	0	0.33
149	reduced calorie	0	0	0	0	0.08	0	1.16	1.15	0	0	0	0	0	0	0.20
150	**Ice cream bar**, chocolate coated	0.01	0	0.06	0	0.17	0.03	7.09	N	N	0	0	0	0	0	0.45
Puddings and chilled desserts																
151	**Cheesecake**, fruit, individual	0	0	0.12	0	0.14	0	0.12	0.12	0	0	0	0	0	0	0.21
152	**Chocolate dairy dessert**, individual	0	0.01	0.06	0	0.12	0.02	2.66	2.70	0.29	2.90	0	0	0	0	0.41
153	**Creme caramel**	0	0	0.01	0	0.03	0	0.30	0.30	0	0	0	0	0	0	0.13
154	**Custard**, ready to eat	0	0	0.05	0.01	0.04	0.01	0.58	0.58	0	0	0	0	0	0	0.07

Milk and milk products

No.	Food	cis n-6					cis n-3					trans
		18:2	18:3	20:3	20:4	22:4	18:3	18:4	20:5	22:5	22:6	Polyunsatd
Yoghurts continued												
142	**Reduced fat yogurt**, frozen	0.04	0	0	0	0	0.01	0	0	0	0	0.01
143	**Virtually fat free/diet yogurt**, fruit, twin pot	0.01	0	0	0	0	0.01	0	0	0	0	0
144	**Fromage frais**, plain	0.15	0	0	0	0	0.02	0	0	0	0	0
145	fruit, children's	0.06	0	0	0	0	0.01	0	0	0	0	0.02
146	**Greek style yogurt**, plain	0.20	0	0	0	0	0.05	0	0	0	0	0.03
147	**Soya, alternative to yogurt**, fruit	0.92	0	0	0	0	0.13	0	0	0	0	0
Ice creams												
148	**Ice cream**, dairy	*0.15*	0.03	0	0	0	0.01	0	0	0	0	0.05
149	reduced calorie	0.06	0.06	0	0	0	0	0	0	0	0	0.06
150	**Ice cream bar**, chocolate coated	1.05	0	0	0	0	0.06	0	0	0	0	0.23
Puddings and chilled desserts												
151	**Cheesecake**, fruit, individual	0.40	0	0	0	0	0.09	0	0	0	0	0.02
152	**Chocolate dairy dessert**, individual	*0.23*	0.02	0	*0.01*	0	0.05	0	0	0.01	0	0.05
153	**Creme caramel**	0.01	0	0	0	0	*0.01*	0	0	0	0	0.04
154	**Custard**, ready to eat	0.05	0	0	0	0	0.01	0	0	0	0	0

Fat and total fatty acids, g per 100g food

No.	Food	Description	Total fat	Satd	cis-Mono unsatd	Polyunsatd Total cis	n-6	n-3	Total trans	Total branched
	Puddings and chilled desserts *continued*									
155	**Fruit fool**, individual	8 samples, 3 brands including strawberry, banana, raspberry, apricot, and gooseberry	10.9	7.26	2.47	0.19	0.26	0.06	0.30	0
156	**Mousse**, chocolate, individual	12 samples, 5 brands	6.5	3.34	1.39	0.09	0.10	0.03	1.27	0.03
157	reduced fat	7 samples, 4 brands	3.7	2.54	0.87	0.09	0.08	0.01	N	N
158	**Trifle**, chocolate, individual	5 samples, 2 brands	15.4	9.40	4.12	0.75	0.69	0.08	0.22	0.05
159	fruit	12 samples, 7 brands including fruit cocktail, strawberry and raspberry; individual and large	9.0	5.42	2.33	0.41	0.37	0.07	0.20	0.44

Saturated fatty acids, g per 100g food

No.	Food	4:0	6:0	8:0	10:0	12:0	14:0	15:0	16:0	17:0	18:0	20:0	22:0	24:0
	Puddings and chilled desserts continued													
155	**Fruit fool**, individual	0.34	0.18	0.11	0.29	0.37	1.20	0.13	3.49	0.06	1.03	0.07	0	0
156	**Mousse**, chocolate, individual	0.02	0.01	0.03	0.04	0.18	0.21	0.03	1.60	0.02	1.19	N	0	0
157	reduced fat	0.01	0.03	0.03	0.03	0.15	0.11	0.01	0.99	0.03	1.11	0.04	0	0
158	**Trifle**, chocolate, individual	0.36	0.23	0.11	0.34	0.40	1.12	0.12	4.61	0.05	1.94	0.12	0	0
159	fruit	0.28	0.13	0.09	0.19	0.30	0.86	0.09	2.55	0.06	0.86	0.01	0	0

Monounsaturated fatty acids, g per 100g food

No.	Food	cis							cis/trans 18:1 n-9	cis/trans 18:1 n-7	cis 20:1	cis 22:1	cis/trans 22:1 n-11	cis/trans 22:1 n-9	cis 24:1	trans Monounsatd
		10:1	12:1	14:1	15:1	16:1	17:1	18:1								
	Puddings and chilled desserts continued															
155	**Fruit fool**, individual	0.03	0.03	0.18	0.03	0.17	0.05	1.97	1.97	0	0.02	0	0	0	0	0.18
156	**Mousse**, chocolate, individual	0	0	0.01	0	0.03	0	1.34	1.34	0	0	0	0	0	0	1.23
157	reduced fat	0	0	0.01	0	0.02	0	0.84	0.84	0	0	0	0	0	0	N
158	**Trifle**, chocolate, individual	0.05	0	0.15	0	0.19	0.03	3.68	3.68	0	0.02	0	0	0	0	0.20
159	fruit	0	0	0.12	0.03	0.14	0.02	1.99	1.99	0	0.03	0	0	0	0	0.18

No.	Food	cis n-6					cis n-3					trans
		18:2	18:3	20:3	20:4	22:4	18:3	18:4	20:5	22:5	22:6	Polyunsatd
Puddings and chilled desserts continued												
155	**Fruit fool**, individual	0.14	0	0	0	0	0.06	0	0	0	0	0.12
156	**Mousse**, chocolate, individual	*0.06*	0	0	0	0	0.03	0	0	0	0	0.04
157	reduced fat	0.08	0	0	0	0	0.01	0	0	0	0	0
158	**Trifle**, chocolate, individual	0.67	0	0	0	0	0.08	0	0	0	0	0.02
159	fruit	0.34	0	0	0	0	0.07	0	0	0	0	0.02

Eggs and egg dishes

Fat and total fatty acids, g per 100g food

No.	Food	Description	Total fat	Satd	cis-Mono unsatd	Polyunsatd Total cis	n-6	n-3	Total trans	Total branched
	Eggs									
160	**Chicken eggs**	12 samples; battery and free range	11.2	3.15	4.31	1.68	1.61	0.08	0.12	0
161	**Duck eggs**	12 samples	11.8	2.83	4.82	2.05	1.81	0.20	0.08	0.07
162	**Quail eggs**	36 samples	11.2	3.07	4.89	1.26	1.11	0.16	0.02	0.07
	Savoury egg dishes									
163	**Egg fried rice**, takeaway	10 samples from different outlets	4.9	0.62	2.34	1.31	1.00	0.31	0	0
164	**Scotch eggs**, retail	2 samples	16.0	4.37	6.60	3.26	2.74	0.58	0.23	0.01

160 to 164

Eggs and egg dishes

Saturated fatty acids, g per 100g food

No.	Food	4:0	6:0	8:0	10:0	12:0	14:0	15:0	16:0	17:0	18:0	20:0	22:0	24:0
Eggs														
160	**Chicken eggs**	0	0	0	0	0	0.03	0	2.23	0.04	0.78	0.05	0.01	0
161	**Duck eggs**	0	0	0	0	0	0.10	0.02	1.50	0.05	1.13	0.03	0	0
162	**Quail eggs**	0	0	0	0	0	0.05	0.01	2.13	0.02	0.86	0.01	0	0
Savoury egg dishes														
163	**Egg fried rice**, takeaway	0	0	0	0	0	0.01	0.01	0.42	Tr	0.12	0.02	0.01	0.02
164	**Scotch eggs**, retail	0	0	0	0	0	0.22	0.03	2.25	0.06	1.73	0.04	0.01	0

Monounsaturated fatty acids, g per 100g food

No.	Food	cis							cis/trans		cis		cis/trans		cis	trans
		10:1	12:1	14:1	15:1	16:1	17:1	18:1	18:1 n-9	18:1 n-7	20:1	22:1	22:1 n-11	22:1 n-9	24:1	Monounsatd
Eggs																
160	**Chicken eggs**	0	0	0	0	0.25	0.04	3.98	3.97	0	0.03	0	0	0	0.01	0.11
161	**Duck eggs**	0	0	0	0.01	0.59	0.02	4.13	3.86	0.27	0.07	0	0	0	0	0.06
162	**Quail eggs**	0	0	0.01	0.01	0.47	0.01	4.38	4.19	0.19	0.01	0	0	0	0	0.01
Savoury egg dishes																
163	**Egg fried rice**, takeaway	0	0	0	0	0.03	0	2.24	N	N	0.06	0.01	Tr	Tr	0	0
164	**Scotch eggs**, retail	0	0	0.03	0.01	0.46	0.04	5.83	5.99	0.01	0.16	0.06	N	N	0	0.17

Eggs and egg dishes

Polyunsaturated fatty acids, g per 100g food

No.	Food	cis n-6						cis n-3						trans
		18:2	18:3	20:3	20:4	22:4	18:3	18:4	20:5	22:5	22:6	Polyunsatd
Eggs												
160	**Chicken eggs**	1.60	0	0	0	0	0.08	0	0	0	0	0.01
161	**Duck eggs** [a]	1.29	0	0.04	0.36	0.04	0.11	0	0	0.02	0.07	0.02
162	**Quail eggs**	0.96	0.01	0	0.13	0	0.05	0	0	0	0.11	0.01
Savoury egg dishes												
163	**Egg fried rice**, takeaway	0.98	0	0	0.01	0.01	0.29	0	0	0.01	0	0
164	**Scotch eggs**, retail [b]	2.63	0	0	0.07	0	*0.41*	0	0.01	0.03	0.07	0.06

[a] Contains 0.03g 16:2, 0.07g 20:2, 0.03 22:3 per 100g food
[b] Contains 0.04g 20:2 per 100g food

Fat and total fatty acids, g per 100g food

No.	Food	Description	Total fat	Satd	cis-Mono unsatd	Polyunsatd Total cis	n-6	n-3	Total trans	Total branched
	Cooking fats									
165	**Compound cooking fat**	Data from TRANSFAIR; 14 samples including White Flora, Trex, Cookeen, White Cap, own brands. Includes standard and polyunsaturated products	99.5	24.52	31.56	27.72	27.06	1.49	10.37	0
166	**Dripping**, beef	Data from TRANSFAIR; 5 samples, different brands	99.0	50.62	34.19	1.86	1.98	0.44	4.42	0.76
167	**Ghee**, vegetable	3 samples, 2 brands	99.4	48.37	36.30	9.31	9.41	0.29	1.05	0
168	**Lard**	10 samples, 3 brands	99.0	40.60	42.99	9.79	9.16	0.51	0	0
169	**Suet**, shredded [a]	Data from TRANSFAIR; 5 samples, different brands	90.5	48.33	27.27	1.59	1.71	0.54	5.66	0.78
170	vegetable [b]	10 samples, 5 brands	87.9	44.86	13.39	11.99	12.52	0.18	10.66	0.11
	Spreading fats [†]									
171	**Butter** [c]	24 samples of UK, Irish, Danish and French butters	82.2	52.09	18.48	2.27	1.41	0.68	2.87	1.80
172	spreadable	8 samples, different brands	82.5	45.38	21.47	1.85	2.55	0.92	2.82	2.84
173	**Blended spread** 70-80%fat	Manufacturer's data (St Ivel)	70.5	16.50	18.70	26.50	26.40	0.20	4.30	Tr
174	40% fat	10 samples including Anchor half fat butter	39.6	18.40	8.29	4.35	3.86	0.49	N	0.53
175	**Margarine**, catering	3 samples, 2 brands	81.7	16.71	31.24	17.50	14.68	4.37	12.50	0
176	hard, vegetable fats only	10 samples, different brands	84.4	37.07	19.92	8.62	8.33	1.29	15.03	0
177	soft, not polyunsaturated	20 samples including Stork Special Blend	80.0	23.38	31.13	12.08	9.76	2.68	8.87	0.82
178	soya	10 samples; 5 own brands	82.7	18.92	17.37	33.41	30.83	3.79	6.04	0.31

[†] The food industry have recently lowered the trans fatty acid content of many spreading fats. Refer to manufacturers for information on the most current fatty acid profiles for individual products

[a] Contains 2.27g unidentified fatty acids per 100g food

[b] Contains 0.07g unidentified fatty acids per 100g food

[c] Contains 0.04g unidentified fatty acids per 100g food

Saturated fatty acids, g per 100g food

No.	Food	4:0	6:0	8:0	10:0	12:0	14:0	15:0	16:0	17:0	18:0	20:0	22:0	24:0
Cooking fats														
165	**Compound cooking fat**	0	0	0.51	0.13	0.14	0.41	0.09	15.65	0.09	6.51	0.64	0.36	0
166	**Dripping**, beef	0	0	0.46	0.21	0.09	3.12	N	24.32	1.50	20.70	0.18	0.03	0
167	**Ghee**, vegetable	0	0	0.19	0	0.19	0.95	0	41.72	0.10	4.75	0.38	0.10	0
168	**Lard**	0	0	0	0	0.02	1.46	0	24.36	0.23	14.35	0.19	0	0
169	**Suet**, shredded	0	0	0.47	0.20	0.09	2.79	1.00	20.72	1.51	21.34	0.16	0.03	0
170	vegetable	0	0	0	0	0.21	0.71	0	27.17	0.10	16.00	0.43	0.23	0
Spreading fats														
171	**Butter** [a]	3.32	1.94	1.09	2.34	2.69	8.37	0.84	22.04	0.56	8.56	0.13	0.05	0.01
172	spreadable	N	1.52	1.14	2.68	3.18	8.46	0.89	19.38	0.69	7.34	0.09	N	0
173	**Blended spread** 70-80%fat	N	N	N	N	N	0.30	0	11.80	0.00	3.50	0.30	0.50	0
174	40% fat	0.83	0.49	0.27	0.53	0.57	1.86	0.19	9.35	0.11	4.13	0.08	0	0
175	**Margarine**, catering	0	0	0	0	0.23	1.17	0.16	7.11	0.16	5.86	0.94	0.94	0.16
176	hard, vegetable fats only	0	0	0.65	0.65	9.10	3.51	0	15.84	0.04	6.56	0.36	0.36	0
177	soft, not polyunsaturated	0	0	0.05	0	0.10	4.33	0.25	12.57	0.25	3.95	1.05	0.82	0
178	soya	0	0	0	0	0.07	0.23	0	12.52	0.09	5.33	0.29	0.27	0.11

[a] Contains 0.05g 11:0, 0.07g 13:0 per 100g food

Monounsaturated fatty acids, g per 100g food

No.	Food	cis							cis/trans 18:1 n-9	18:1 n-7	cis 20:1	cis 22:1	cis/trans 22:1 n-11	22:1 n-9	cis 24:1	trans Monounsatd
		10:1	12:1	14:1	15:1	16:1	17:1	18:1								

Cooking fats

No.	Food	10:1	12:1	14:1	15:1	16:1	17:1	18:1	18:1 n-9	18:1 n-7	20:1	22:1	22:1 n-11	22:1 n-9	24:1	trans Monounsatd
165	**Compound cooking fat**	0	0	0	0	0.10	0	30.88	23.87	7.01	0.49	0.09	N	N	0	9.53
166	**Dripping**, beef	0	0	1.03	0	2.07	0	30.86	24.67	6.19	0.23	0	0	0	0	3.85
167	**Ghee**, vegetable	0	0	0	0	0.19	0	36.02	35.35	0.67	0.10	0	0	0	0	0.67
168	**Lard**	0	0	0	0	2.44	0.23	39.39	32.94	3.98	0.93	0	0	0	0	0
169	**Suet**, shredded	0	0	0	0	1.51	0	25.59	20.44	3.76	0.17	0	0	0	0	5.00
170	vegetable	0	0	0	0	0.05	0	13.26	13.26	2.95	0.07	0	0	0	0	9.96

Spreading fats

No.	Food	10:1	12:1	14:1	15:1	16:1	17:1	18:1	18:1 n-9	18:1 n-7	20:1	22:1	22:1 n-11	22:1 n-9	24:1	trans Monounsatd
171	**Butter**	0.23	0.06	0.69	0.01	1.24	0.21	15.80	14.73	2.41	0.17	0	0	0	0.08	2.43
172	spreadable	0	0	0	0	1.33	0.30	18.78	18.78	3.51	1.02	0	0	0	0.04	1.20
173	**Blended spread** 70-80%fat	0	N	N	0	N	0	18.70	N	N	0	0	0	0	0	4.10
174	40% fat	0	0.19	0	0	0.30	0	7.27	5.75	7.16	0.08	0.45	0	0.45	0	6.28
175	**Margarine**, catering	0	0	0	0	0.23	0.31	29.21	25.93	3.28	1.09	0.31	N	N	0.08	10.93
176	hard, vegetable fats only	0	0	0	0	0.16	0.02	19.21	18.93	12.93	0.36	0.16	0	0.26	0	14.02
177	soft, not polyunsaturated	0	0	0.03	0	1.50	0.15	24.65	21.67	7.21	2.73	2.06	0.51	1.56	0	8.51
178	soya	0	0	0	0	0.08	0	17.16	17.16	2.83	0.14	0	0	0	0	4.83

No.	Food	18:2	cis n-6				cis n-3					trans
			18:3	20:3	20:4	22:4	18:3	18:4	20:5	22:5	22:6	Polyunsatd
Cooking fats												
165	**Compound cooking fat**	26.43	0	0	0	0	1.22	0	0.06	0	0	0.84
166	**Dripping**, beef	1.46	0	0	0	0	0.40	0	0	0	0	0.57
167	**Ghee**, vegetable	9.12	0	0	0	0	0.19	0	0	0	0	0.38
168	**Lard** a	8.54	0.06	0.17	0.11	0	0.51	0	0	0	0	0
169	**Suet**, shredded	1.08	0	0.02	0	0	0.49	0	0.02	0	0	0.66
170	vegetable	11.81	0	0	0	0	0.18	0	0	0	0	0.70
Spreading fats												
171	**Butter** b	0.95	0.02	0.05	0.09	0	0.46	0.02	0.06	0.08	0	0.56
172	spreadable c	0.93	N	N	N	0	0.92	N	N	N	0	1.62
173	Blended spread 70-80% fat	26.30	0	0	0	0	0.20	0	0	0	0	0.20
174	40% fat	3.86	0	0	0	0	0.49	0	0	0	0	N
175	**Margarine**, catering d	12.42	0	0	0	0	3.36	0	0	0	0	1.56
176	hard, vegetable fats only	7.34	0	0	0	0	1.27	0	0	0	0	1.01
177	soft, not polyunsaturated e	9.48	0	0	0	0	2.44	0	0	0	0	0.36
178	soya f	29.90	0	0	0	0	3.51	0	0	0	0	1.21

a Contains 0.40g 20:2 per 100g food
b Contains 0.03g 21:5 per 100g food
c Contains 0.32g 20 poly, 0.39g 22poly per 100g food

d Contains 0.70g 20:2, 1.02g 22:2 per 100g food
e Contains 0.15g 20:2 per 100g food
f Contains 0.14g 20 poly per 100g food

Fats and oils

Fat and total fatty acids, g per 100g food

No.	Food	Description	Total fat	Satd	cis-Mono unsatd	Polyunsatd Total cis	n-6	n-3	Total trans	Total branched
	Spreading fats † *continued*									
179	**Fat spread** 70–80% fat, not polyunsaturated ab	Data from TRANSFAIR; 5 samples including Clover	75.5	20.93	30.29	8.97	7.25	2.54	11.45	0.06
180	70% fat, monounsaturated cd	Data from TRANSFAIR; 5 samples including Utterly Butterly	70.0	9.44	31.63	14.84	11.73	3.79	10.45	0.04
181	70% fat, polyunsaturated ef	Data from TRANSFAIR; 5 samples including Vitalite	68.5	16.23	15.12	33.42	33.40	0.16	0.25	0.01
182	60% fat, with olive oil	5 samples including Olivio	62.7	11.27	31.17	11.63	10.13	2.34	5.99	Tr
183	40% fat, not polyunsaturated gh	Data from TRANSFAIR; 5 samples including Gold Light	37.5	8.43	16.85	5.84	4.45	1.69	4.43	0.02
184	35–40% fat, polyunsaturated ij	Data from TRANSFAIR; 5 samples including Vitalite Light	41.5	7.33	9.26	19.48	19.58	0.20	3.27	0.01
185	20–25% fat, not polyunsaturated kl	Data from TRANSFAIR; 5 samples including Gold Lowest	25.5	6.82	10.32	3.15	2.66	0.70	3.90	0.01
	Oils									
186	**Blackcurrant seed oil**	9 samples	99.9	8.20	11.30	75.16	60.04	12.45	Tr	0
187	**Borage oil**	12 samples	99.9	14.41	24.68	56.24	55.91	0.33	Tr	0
188	**Coconut oil**	35 samples	99.9	86.50	6.00	1.50	1.50	0	Tr	0
189	**Cod liver oil**	20 samples	99.9	21.10	44.60	30.50	3.50	24.40	Tr	0

† The food industry have recently lowered the trans fatty acid content of many spreading fats. Refer to manufacturers for information on the most current fatty acid profiles for individual products

a Contains 0.48g unidentified fatty acids per 100g food   b Krona contains 70.0g fat, 32.28g satd, 24.49g cis n-9 monounsatd, 7.36g cis n-6 and 1.23g cis n-3 polyunsatd, 0.58g trans per 100g food

c Contains 0.51g unidentified fatty acids per 100g food   d I Can't Believe It's Not Butte:I contains 70.0g fat, 22.0g satd, 23.55g cis-monounsatd, 19.8g cis-polyunsatd, 0.6g trans per 100g food

e Contains 0.45g unidentified fatty acids per 100g food   f Flora contains 70.0g fat, 15.5g satd, 20.91g cis n-9 monounsatd, 31.53g cis n-6 and 1.41g cis n-3 polyunsatd, 0.64g trans per 100g food

g Contains 0.27g unidentified fatty acids per 100g food   h Delight Low Fat contains 38 0g fat, 12.67g satd, 18.0g cis n-9 monounsatd, 5.34g cis n-6 and 1.78g cis n-3 polyunsatd, 0.60g trans per 100g food

i Contains 0.33g unidentified fatty acids per 100g food   j Flora Light contains 38.0g fat, 8.86g satd, 10.62g cis n-9 monounsatd, 17.44g cis n-6 and 0.62g cis n-3 polyunsatd, 0.43g trans per 100g food

k Contains 0.18g unidentified fatty acids per 100g food   l Delight Diet contains 23g fat 7.52g satd, 10.41g cis n-9 monounsatd, 3.16g cis n-6 and 1.05g cis n-3 polyunsatd, 0.35g trans per 100g food

Saturated fatty acids, g per 100g food

No.	Food	4:0	6:0	8:0	10:0	12:0	14:0	15:0	16:0	17:0	18:0	20:0	22:0	24:0
	Spreading fats continued													
179	**Fat spread** 70-80% fat, not polyunsaturated	0	0	0.95	0.67	2.74	1.72	0.19	9.64	0.12	4.33	0.40	0.17	0
180	70% fat, monounsaturated	0	0	0.35	0.11	0.07	0.10	0.05	5.23	0.05	2.74	0.54	0.20	0
181	70% fat, polyunsaturated	0	0	0.56	0.28	1.11	0.60	0.06	10.12	0.06	2.74	0.42	0.31	0
182	60% fat, with olive oil	0	0	0	0	0.12	0.18	0	7.07	0.06	3.30	0.30	0.18	0.06
183	40% fat, not polyunsaturated	0	0	0.19	0	0.04	0.13	0.03	5.93	0.02	1.70	0.24	0.09	0
184	35-40% fat, polyunsaturated	0	0	0.22	0.06	0.03	0.07	0.03	4.33	0.02	2.07	0.30	0.21	0
185	20-25% fat, not polyunsaturated	0	0	0.14	0.06	0.02	0.12	0.03	4.71	0.03	1.50	0.16	0.05	0
	Oils													
186	**Blackcurrant seed oil**	0	0	0	0	0	0.10	0	6.39	0.08	1.38	0.15	0.11	0
187	**Borage oil**	0	0	0	0	0	0.07	0	9.83	0.06	4.05	0.25	0.14	0
188	**Coconut oil**	0	0	6.90	6.20	45.00	17.00	0	8.40	0	2.50	0.10	0	0
189	**Cod liver oil**	0	0	0	0	0	5.60	Tr	12.50	0	3.00	Tr	0	0

Monounsaturated fatty acids, g per 100g food

No.	Food	cis							cis/trans		cis		cis/trans		cis	trans
		10:1	12:1	14:1	15:1	16:1	17:1	18:1	18:1 n-9	18:1 n-7	20:1	22:1	22:1 n-11	22:1 n-9	24:1	Monounsatd
	Spreading fats continued															
179	**Fat spread** 70-80% fat, not polyunsaturated	0	0	0	0	0.17	0	29.40	22.35	6.81	0.65	0.06	N	N	0	10.63
180	70% fat, monounsaturated	0	0	0	0	0.11	0	30.80	22.32	8.29	0.63	0.09	N	N	0	9.76
181	70% fat, polyunsaturated	0	0	0	0	0.09	0	14.96	12.40	2.56	0.07	0	0	0	0	0.11
182	60% fat, with olive oil	0	0	0	0	0.18	0.06	30.51	28.11	2.40	0.30	0.06	N	N	0.06	5.15
183	40% fat, not polyunsaturated	0	0	0	0	0.04	0	16.48	12.22	4.20	0.29	0.03	N	N	0	4.12
184	35-40% fat, polyunsaturated	0	0	0	0	0.02	0	9.18	6.07	3.04	0.06	0	0	0	0	2.97
185	20-25% fat, not polyunsaturated	0	0	0	0	0.05	0	10.11	7.51	2.56	0.15	0.01	0	0.01	0	3.69
	Oils															
186	**Blackcurrant seed oil**	0	0	0	0	0.19	0	10.10	10.10	0	0.93	0.07	N	N	0	Tr
187	**Borage oil**	0	0	0	0	0.29	0.02	17.24	17.24	0	3.78	2.24	N	N	1.11	Tr
188	**Coconut oil**	0	0	0	0	0	0	6.00	6.00	0	0	0	0	0	0	Tr
189	**Cod liver oil**	0	0	0	0	7.10	0	18.30	13.17	5.13	11.10	8.10	8.10	0	0	Tr

Fats and oils

Polyunsaturated fatty acids, g per 100g food

No.	Food	cis n-6					cis n-3					trans
		18:2	18:3	20:3	20:4	22:4	18:3	18:4	20:5	22:5	22:6	Polyunsatd
Spreading fats continued												
179	**Fat spread** 70-80% fat, not polyunsaturated [a]	6.78	0	0	0	0	2.14	0	0.03	0	0	0.82
180	70% fat, monounsaturated [b]	11.23	0	0	0	0	3.51	0	0.03	0	0	0.69
181	70% fat, polyunsaturated	33.26	0	0	0	0	0.09	0	0.07	0	0	0.14
182	60% fat, with olive oil	9.65	0	0	0	0	1.98	0	0	0	0	0.84
183	40% fat, not polyunsaturated [c]	4.27	0	0	0	0	1.54	0	0.01	0	0	0.31
184	35-40% fat, polyunsaturated	19.30	0	0	0	0	0.15	0	0.03	0	0	0.30
185	20-25% fat, not polyunsaturated	2.54	0	0	0	0	0.60	0	Tr	0	0	0.21
Oils												
186	**Blackcurrant seed oil**	45.41	14.63	0	0	0	12.45	2.67	0	0	0	Tr
187	**Borage oil**	35.27	20.64	0	0	0	0.33	0	0	0	0	Tr
188	**Coconut oil**	1.50	0	0	0	0	0	0	0	0	0	Tr
189	**Cod liver oil** [d]	2.60	0	0	0.90	0	1.10	2.10	10.80	1.40	8.30	Tr

[a] Contains 0.01g 20:2, 0.03g 22:3 per 100g food
[b] Contains 0.03g 20:2, 0.05g 22:3 per 100g food
[c] Contains 0.02g 22:3 per 100g food
[d] Contains 0.60g 16:2, 0.40g 16:3, 1.00g 16:4, 0.70g 21:5 per 100g food

Fat and total fatty acids, g per 100g food

No.	Food	Description	Total fat	Satd	cis-Mono unsatd	Polyunsatd Total cis	n-6	n-3	Total trans	Total branched
	Oils continued									
190	**Corn oil**	42 samples	99.9	14.50	29.90	51.30	50.40	0.90	Tr	0
191	**Cottonseed oil**	55 samples	99.9	26.10	18.30	50.20	50.10	0.10	Tr	0
192	**Evening primrose oil**	18 samples	99.9	7.70	10.63	77.05	76.93	0.12	Tr	0
193	**Grapeseed oil**	25 samples	99.9	10.91	18.91	65.71	65.42	0.29	Tr	0
194	**Hazelnut oil**	10 samples	99.9	7.80	76.50	11.20	11.10	0.10	Tr	0
195	**Olive oil**	35 samples including virgin and extra virgin	99.9	14.30	73.00	8.20	7.50	0.70	Tr	0
196	**Palm oil**	55 samples	99.9	47.80	37.10	10.40	10.10	0.30	Tr	0
197	**Peanut oil**	71 samples	99.9	20.00	44.40	31.00	31.00	0	Tr	0
198	**Rapeseed oil**	100 samples	99.9	6.60	59.20	29.30	19.70	9.60	Tr	0
199	**Safflower oil**	28 samples	99.9	9.70	11.90	74.00	73.90	0.10	Tr	0
200	**Sesame oil**	22 samples and literature sources	99.9	14.60	37.50	43.40	43.10	0.30	Tr	0
201	**Soya oil**	39 samples	99.9	15.60	21.20	58.80	51.50	7.30	Tr	0
202	**Sunflower oil**	46 samples	99.9	12.00	20.50	63.30	63.20	0.10	Tr	0
203	**Walnut oil**	13 samples	99.9	9.10	16.40	69.90	58.40	11.50	Tr	0
204	**Wheatgerm oil**	35 samples	99.9	18.50	16.70	60.40	55.10	5.30	Tr	0
205	**Vegetable oil**, blended	Data from Institute of Human Nutrition and Brain Chemistry	99.9	11.67	53.20	29.76	23.26	6.50	Tr	0

Fats and oils

Saturated fatty acids, g per 100g food

No.	Food	4:0	6:0	8:0	10:0	12:0	14:0	15:0	16:0	17:0	18:0	20:0	22:0	24:0
	Oils continued													
190	**Corn oil**	0	0	0	0	0.10	0.10	0	11.30	0	2.10	0.50	0.20	0.20
191	**Cottonseed oil**	0	0	0	0	Tr	0.80	0	22.40	0	2.50	0.30	0.10	0
192	**Evening primrose oil**	0	0	0	0	0	0	0	5.64	0.09	1.68	0.23	0.05	0
193	**Grapeseed oil**	0	0	0	0	0	0.06	0	6.85	0.10	3.90	0	0	0
194	**Hazelnut oil**	0	0	0	0	0	Tr	0	5.30	0.10	2.30	0.10	0	0
195	**Olive oil**	0	0	0	0	0.10	0.10	0	10.10	0.10	3.00	0.40	0.10	0.40
196	**Palm oil**	0	0	0	0	0.10	1.00	0	41.80	0	4.60	0.30	0	0
197	**Peanut oil**	0	0	0	0	Tr	Tr	0	10.90	0	3.20	1.30	3.20	1.40
198	**Rapeseed oil**	0	0	0	0	0	Tr	0	4.20	Tr	1.50	0.60	0.30	Tr
199	**Safflower oil**	0	0	0	0	0	0.10	0	6.60	0	2.30	0.30	0.30	0.10
200	**Sesame oil**	0	0	0	0	0	Tr	0	8.60	0.10	5.10	0.60	0.10	0.10
201	**Soya oil**	0	0	0	0	0	0.10	0	10.70	0	3.80	0.40	0.50	0.10
202	**Sunflower oil**	0	0	0	0	0	0.10	0	6.20	0	4.30	0.30	0.80	0.30
203	**Walnut oil**	0	0	0	0	0	Tr	0	6.50	0.10	2.40	0.10	Tr	0
204	**Wheatgerm oil**	0	0	0	0	Tr	Tr	0	17.20	0.10	0.90	0.10	Tr	0.10
205	**Vegetable oil**, blended	0	0	0	0	0.02	0.08	0	5.33	0.07	5.00	0.63	0.39	0.15

Fats and oils

Monounsaturated fatty acids, g per 100g food

Oils continued

No.	Food	cis 10:1	12:1	14:1	15:1	16:1	17:1	18:1	cis/trans 18:1 n-9	18:1 n-7	cis 20:1	22:1	cis/trans 22:1 n-11	22:1 n-9	cis 24:1	trans Monounsatd
190	Corn oil	0	0	0	0	0.20	0	29.40	29.40	0	0.30	Tr	0	0	0	Tr
191	Cottonseed oil	0	0	0	0	0.80	0	17.40	17.40	0	0.10	Tr	0	0	0	Tr
192	Evening primrose oil	0	0	0	0	0.03	0	10.43	10.43	0	0.14	0.03	0.03	0	0	Tr
193	Grapeseed oil	0	0	0	0	0.10	0	18.81	18.81	0	0	0	0	0	0	Tr
194	Hazelnut oil	0	0	0	0	0.20	0.10	76.10	76.10	0	0.10	0	0	0	0	Tr
195	Olive oil	0	0	0	0	0.70	0.10	71.90	71.90	0	0.30	0	0	0	0	Tr
196	Palm oil	0	0	0	0	Tr	0	37.10	37.10	0	0	0	0	0	0	Tr
197	Peanut oil	0	0	0	0	Tr	0	43.30	43.30	0	1.00	0.10	0	0.10	0	Tr
198	Rapeseed oil	0	0	0	0	0.20	Tr	57.60	57.60	0	1.20	0.20	0	0.20	0	Tr
199	Safflower oil	0	0	0	0	0.10	0	11.40	11.40	0	0.20	0.20	N	N	0	Tr
200	Sesame oil	0	0	0	0	0.10	Tr	37.20	37.20	0	0.20	0	0	0	0	Tr
201	Soya oil	0	0	0	0	0.10	0	20.80	20.80	0	0.20	0.10	0	0.10	0	Tr
202	Sunflower oil	0	0	0	0	0.10	0	20.20	20.20	0	0.10	0.10	0	0.10	0	Tr
203	Walnut oil	0	0	0.10	0	0	0.10	16.20	16.20	0	0	0	0	0	0	Tr
204	Wheatgerm oil	0	0	0	0	0.20	Tr	15.40	15.40	0	1.00	0.10	0	0.10	0	Tr
205	Vegetable oil, blended	0	0	0	0	0.27	0	51.00	51.00	0	1.31	0.46	0	0.46	0.10	Tr

Polyunsaturated fatty acids, g per 100g food

No.	Food	cis n-6					cis n-3					trans
		18:2	18:3	20:3	20:4	22:4	18:3	18:4	20:5	22:5	22:6	Polyunsatd
Oils *continued*												
190	**Corn oil**	50.40	0	0	0	0	0.90	0	0	0	0	Tr
191	**Cottonseed oil**	50.10	0	0	0	0	0.10	0	0	0	0	Tr
192	**Evening primrose oil**	68.81	8.12	0	0	0	0.12	0	0	0	0	Tr
193	**Grapeseed oil**	65.42	0	0	0	0	0.29	0	0	0	0	Tr
194	**Hazelnut oil**	11.10	0	0	0	0	0.10	0	0	0	0	Tr
195	**Olive oil**	7.50	0	0	0	0	0.70	0	0	0	0	Tr
196	**Palm oil**	10.10	0	0	0	0	0.30	0	0	0	0	Tr
197	**Peanut oil**	31.00	0	0	0	0	0	0	0	0	0	Tr
198	**Rapeseed oil**	19.70	0	0	0	0	9.60	0	0	0	0	Tr
199	**Safflower oil**	73.90	0	0	0	0	0.10	0	0	0	0	Tr
200	**Sesame oil**	43.10	0	0	0	0	0.30	0	0	0	0	Tr
201	**Soya oil**	51.50	0	0	0	0	7.30	0	0	0	0	Tr
202	**Sunflower oil**	63.20	0	0	0	0	0.10	0	0	0	0	Tr
203	**Walnut oil**	58.40	0	0	0	0	11.50	0	0	0	0	Tr
204	**Wheatgerm oil**	55.10	0	0	0	0	5.30	0	0	0	0	Tr
205	**Vegetable oil**, blended	23.20	0.06	0	0	0	6.50	0	0	0	0	Tr

Fat and total fatty acids, g per 100g food

No.	Food	Description	Total fat	Satd	cis-Mono unsatd	Total cis	Polyunsatd n-6	n-3	Total trans	Total branched
Beef										
206	**Beef**, average, trimmed lean, *raw*	Average from 6 different cuts	4.3	1.74	1.76	0.20	0.17	0.07	0.14	0.08
207	trimmed fat, *raw*	Average from 3 different cuts	43.0	19.84	17.30	0.62	0.93	0.27	1.99	0.99
208	lean only, *cooked*	Average from 11 different cuts	8.2	3.26	3.41	0.38	0.36	0.09	0.28	0.28
209	fat only, *cooked*	Average from 4 different cuts	45.5	19.73	18.87	1.43	1.61	0.29	2.09	0.95
210	**Braising steak**, *braised*, lean and fat	Calculated from 90% lean and 9% fat	12.7	5.29	4.91	0.71	0.74	0.07	0.51	0.22
211	**Brisket**, *boiled*, lean only	10 samples	11.0	4.36	4.37	0.31	0.32	0.13	0.50	0.21
212	**Fillet steak**, *cooked from steakhouse*, lean only	7 samples	7.0	3.04	2.54	0.36	0.29	0.11	0.20	0.15
213	**Flank**, boneless, *pot roasted*, lean and fat	Calculated from 76% lean and 21% fat	22.3	8.67	9.43	0.55	0.67	0.19	0.85	0.43
214	**Minced beef**, *raw*	10 samples	16.2	6.94	6.30	0.30	0.36	0.12	0.81	0.34
215	*stewed*	10 samples	13.5	5.72	5.21	0.38	0.40	0.16	0.68	0.28
216	extra lean, *raw*	10 samples	9.6	4.02	3.58	0.25	0.27	0.10	0.46	0.20
217	**Rump steak**, *fried in corn oil*, lean only	10 samples	6.6	2.41	2.38	0.84	0.81	0.09	0.20	0.10
218	**Silverside**, *pot roasted*, lean and fat only	Calculated from 80% lean and 18% fat	13.7	5.39	6.02	0.54	0.54	0.12	0.40	0.24
219	**Sirloin steak**, *grilled medium rare*, lean only	19 samples	7.7	3.30	3.03	0.26	0.19	0.07	0.17	0.14
220	**Stewing steak**, *stewed*, lean and fat	Calculated from 84% lean and 14% fat	9.6	3.42	3.82	1.04	0.99	0.09	0.27	0.16
221	**Topside**, *roasted medium-rare*, lean and fat	Calculated from 87% lean and 12% fat	11.4	4.57	4.81	0.43	0.44	0.11	0.44	0.21
Veal										
222	**Veal escalope**, *fried in corn oil*, lean only	9 samples	6.8	1.76	2.44	1.86	1.83	0.07	0.08	0.04
223	**Veal**, minced, *stewed*	5 samples	11.1	4.66	4.22	0.68	0.71	0.06	0.44	0.11

Saturated fatty acids, g per 100g food

No.	Food	4:0	6:0	8:0	10:0	12:0	14:0	15:0	16:0	17:0	18:0	20:0	22:0	24:0
Beef														
206	**Beef**, average, trimmed lean, *raw*	0	0	0	0	0	0.10	0.02	0.97	0.04	0.59	0	0	0
207	trimmed fat, *raw*	0	0	0	0.02	0.03	1.29	0.28	10.54	0.50	7.11	0.06	0	0
208	lean only, *cooked*	0	0	0	0	0.01	0.21	0.04	1.85	0.08	1.07	0.01	0	0
209	fat only, *cooked*	0	0	0	0.02	0.07	1.38	0.27	10.94	0.51	6.49	0.06	0.02	0
210	**Braising steak**, *braised*, lean and fat	0	0	0	0.01	0.03	0.31	0.07	2.80	0.12	1.91	0.02	0	0
211	**Brisket**, *boiled*, lean only	0	0	0	0.01	0.03	0.39	0.06	2.44	0.10	1.33	0.01	0	0
212	**Fillet steak**, *cooked from steakhouse*, lean only	0	0	0	0	0	0.16	0.04	1.59	0.09	1.15	0.01	0	0
213	**Flank**, boneless, *pot roasted*, lean and fat	0	0	0	0.01	0.01	0.64	0.13	5.04	0.23	2.58	0.03	0	0
214	**Minced beef**, *raw*	0	0	0	0.01	0.01	0.46	0.10	3.72	0.17	2.46	0.02	0	0
215	*stewed*	0	0	0	0.01	0.01	0.37	0.08	3.05	0.14	2.05	0.02	0	0
216	extra lean, *raw*	0	0	0	0	0.01	0.25	0.05	2.15	0.10	1.44	0.01	0	0
217	**Rump steak**, *fried in corn oil*, lean only	0	0	0	0	0	0.13	0.03	1.36	0.05	0.82	0.01	0	0
218	**Silverside**, *pot roasted*, lean and fat only	0	0	0	0.01	0.01	0.35	0.07	3.18	0.12	1.64	0.01	0	0
219	**Sirloin steak**, *grilled medium rare*, lean only	0	0	0	0	0	0.20	0.04	1.95	0.08	1.03	0.01	0	0
220	**Stewing steak**, *stewed*, lean and fat	0	0	0	0	0	0.19	0.04	1.90	0.09	1.17	0.01	0	0
221	**Topside**, *roasted medium-rare*, lean and fat	0	0	0	0	0.01	0.26	0.06	2.56	0.12	1.55	0.01	0	0
Veal														
222	**Veal escalope**, *fried in corn oil*, lean only	0	0	0	0	0.01	0.15	0.01	1.10	0.02	0.45	0.01	0.01	0
223	**Veal**, minced, *stewed*	0	0	0	0.01	0.06	0.61	0.03	2.64	0.07	1.22	0.02	0	0

Meat, poultry and game

Monounsaturated fatty acids, g per 100g food

No.	Food	cis 10:1	12:1	14:1	15:1	16:1	17:1	18:1	cis/trans 18:1 n-9	cis/trans 18:1 n-7	cis 20:1	cis 22:1	cis/trans 22:1 n-11	cis/trans 22:1 n-9	cis 24:1	trans Monounsatd
Beef																
206	**Beef**, average, trimmed lean, *raw*	0	0	0.02	0	0.15	0.04	1.54	1.47	0.13	0.01	0	0	0	0	0.11
207	trimmed fat, *raw*	0	0	0.31	0.01	1.59	0.35	14.98	14.34	1.57	0.07	0	0	0	0	1.42
208	lean only, *cooked*	0	0	0.05	0	0.30	0.07	2.98	2.86	0.25	0	0	0	0	0	0.21
209	fat only, *cooked*	0	0	0.40	0.01	1.89	0.40	16.04	15.62	1.50	0.14	0	0	0	0	1.62
210	**Braising steak**, *braised*, lean and fat	0	0	0.06	0.01	0.36	0.09	4.33	4.25	0.42	0.03	0.02	0	0.02	0	0.41
211	**Brisket**, *boiled*, lean only	0	0	0	0	0.47	0	3.90	3.72	0.43	0	0	0	0	0	0.37
212	**Fillet steak**, *cooked from steakhouse*, lean only	0	0	0	0	0.19	0	2.36	2.25	0.21	0	0	0	0	0	0.16
213	**Flank**, boneless, *pot roasted*, lean and fat	0	0	0.22	0	0.98	0.22	8.02	7.65	0.72	0	0	0	0	0	0.62
214	**Minced beef**, *raw*	0	0	0	0	0.61	0	5.69	5.44	0.68	0	0	0	0	0	0.64
215	*stewed*	0	0	0	0	0.51	0	4.70	4.52	0.53	0	0	0	0	0	0.50
216	extra lean, *raw*	0	0	0	0	0.33	0	3.24	3.10	0.38	0	0	0	0	0	0.34
217	**Rump steak**, *fried in corn oil*, lean only	0	0	0	0	0.20	0	2.18	2.09	0.18	0	0	0	0	0	0.14
218	**Silverside**, *pot roasted*, lean and fat only	0	0	0.12	0	0.62	0.13	5.14	4.88	0.39	0	0	0	0	0	0.28
219	**Sirloin steak**, *grilled medium rare*, lean only	0	0	0	0	0.28	0	2.75	2.66	0.18	0	0	0	0	0	0.17
220	**Stewing steak**, *stewed*, lean and fat	0	0	0.05	0	0.31	0.07	3.39	3.23	0.29	0	0	0	0	0	0.21
221	**Topside**, *roasted medium-rare*, lean and fat	0	0	0.07	0	0.43	0.10	4.21	4.03	0.37	0	0	0	0	0	0.32
Veal																
222	**Veal escalope**, *fried in corn oil*, lean only	0	0	0	0	0.17	0	2.27	2.16	0.12	0	0	0	0	0	0.04
223	**Veal**, minced, *stewed*	0	0	0	0	0.40	0	3.83	3.67	0.36	0	0	0	0	0	0.36

No.	Food	cis n-6					cis n-3					trans
		18:2	18:3	20:3	20:4	22:4	18:3	18:4	20:5	22:5	22:6	Polyunsatd
Beef												
206	**Beef**, average, trimmed lean, *raw*	0.11	0	0.01	0.02	0	0.03	0	0.01	0.02	0	0.03
207	trimmed fat, *raw*	0.41	0	0	0	0	0.19	0.01	0	0	0	0.57
208	lean only, *cooked*	0.26	0	0.01	0.03	0	0.05	0	0.01	0.01	0	0.07
209	fat only, *cooked*	1.19	0	0.01	0.01	0	0.20	0.01	0.02	0	0	0.47
210	**Braising steak**, *braised*, lean and fat	0.63	0	0	0.02	0	0.05	0	0	0	0	0.10
211	**Brisket**, *boiled*, lean only	0.16	0	0.01	0.04	0	0.07	0	0.01	0.02	0	0.13
212	**Fillet steak**, *cooked from steakhouse*, lean only	0.20	0	0.01	0.05	0	0.07	0	0.02	0.02	0	0.04
213	**Flank**, boneless, *pot roasted*, lean and fat	0.45	0	0	0.03	0	0.10	0	0	0.03	0	0.23
214	**Minced beef**, *raw*	0.18	0	0.01	0	0	0.09	0.01	0	0	0	0.18
215	*stewed*	0.21	0	0.01	0.03	0	0.08	0.01	0.01	0.03	0	0.18
216	extra lean, *raw*	0.14	0	0.01	0.02	0	0.06	0	0	0.02	0	0.12
217	**Rump steak**, *fried in corn oil*, lean only	0.71	0	0.01	0.05	0	0.05	0	0.02	0	0	0.06
218	**Silverside**, *pot roasted*, lean and fat only	0.37	0	0.02	0.05	0	0.07	0	0.01	0.02	0	0.12
219	**Sirloin steak**, *grilled medium rare*, lean only	0.17	0	0.01	0.02	0	0.05	0	0.01	0.01	0	0
220	**Stewing steak**, *stewed*, lean and fat	0.91	0	0.01	0.02	0	0.06	0	0.01	0.02	0	0.06
221	**Topside**, *roasted medium-rare*, lean and fat	0.26	0	0.01	0.04	0	0.07	0	0.02	0.02	0	0.12
Veal												
222	**Veal escalope**, *fried in corn oil*, lean only	1.70	0	0.02	0.08	0	0.04	0	0.01	0.02	0	0.04
223	**Veal**, minced, *stewed* [a]	0.58	0	0.01	0.04	0	0.04	0	0	0	0	0.09

[a] Contains 0.01g 20:2 per 100g food

No.	Food	Description	Total fat	Satd	cis-Mono unsatd	Polyunsatd			Total trans	Total branched
						Total cis	n-6	n-3		
	Lamb									
224	**Lamb**, average, trimmed lean, *raw*	Average from 9 different cuts	8.0	3.46	2.58	0.36	0.28	0.16	0.60	0.15
225	trimmed fat, *raw*	Average from 9 different cuts	52.8	25.53	15.89	1.68	1.59	0.76	4.89	1.17
226	lean only, *cooked*	Average from 16 different cuts	12.5	5.36	4.06	0.59	0.48	0.23	0.93	0.24
227	fat only, *cooked*	Average from 6 different cuts	56.3	26.67	17.53	1.79	1.38	0.82	5.19	1.22
228	**Best end neck cutlets**, *barbecued*, lean only	10 samples	13.9	6.26	4.24	0.54	0.47	0.25	1.10	0.26
229	**Chump steaks/chops**, *fried in corn oil*, lean only	35 samples of a mixture of chops and steaks	11.2	4.58	3.47	0.86	0.71	0.23	0.71	0.40
230	**Leg**, average, *raw*, lean and fat	Calculated from 83% lean and 17% fat	12.3	5.36	4.05	0.63	0.47	0.27	0.93	0.22
231	**Leg, whole**, *roasted medium*, lean and fat	Calculated from 89% lean and 11% fat	14.2	5.66	5.09	0.66	0.56	0.29	1.15	0.27
232	**Loin chops**, *grilled*, lean only	33 samples	10.7	4.64	3.30	0.51	0.41	0.21	0.85	0.22
233	**Minced lamb**, *stewed*, lean and fat	10 samples	12.3	5.68	4.12	0.49	0.36	0.21	0.89	0.26
234	**Neck fillet**, strips, *stir-fried in corn oil*, lean only	10 samples	20.0	7.84	6.43	2.02	1.86	0.32	1.30	0.33
235	**Rack of lamb**, *roasted*, lean and fat	Calculated from 66% lean and 34% fat	30.1	14.20	9.26	1.10	0.90	0.45	2.29	0.64
236	**Shoulder**, half bladeside, *pot-roasted*, lean and fat	Calculated from 72% lean and 28% fat	25.6	11.46	8.14	1.10	0.93	0.43	2.08	0.51
237	whole, *roasted*, lean and fat	Calculated from 78% lean and 22% fat	22.1	9.99	7.17	0.86	0.73	0.32	1.70	0.43
238	**Stewing lamb**, *stewed*, lean and fat	Calculated from 85% lean and 15% fat	20.1	8.81	6.40	1.17	0.99	0.35	1.49	0.39

Meat, poultry and game

Saturated fatty acids, g per 100g food   *224 to 238*

Lamb

No.	Food	4:0	6:0	8:0	10:0	12:0	14:0	15:0	16:0	17:0	18:0	20:0	22:0	24:0
224	**Lamb**, average, trimmed lean, *raw*	0	0	0	0.02	0.04	0.38	0.05	1.59	0.08	1.29	0.01	0	0
225	trimmed fat, *raw*	0	0	0	0.16	0.32	2.95	0.41	10.96	0.66	10.00	0.08	0	0
226	lean only, *cooked*	0	0	0	0.03	0.06	0.53	0.08	2.47	0.14	2.05	0.02	0	0
227	fat only, *cooked*	0	0	0	0.15	0.30	2.89	0.42	11.64	0.74	10.46	0.07	0	0
228	**Best end neck cutlets**, *barbecued*, lean only	0	0	0	0.04	0.08	0.74	0.09	2.73	0.14	2.42	0.02	0	0
229	**Chump steaks/chops**, *fried in corn oil*, lean only	0	0	0	0.02	0.05	0.46	0.06	2.07	0.11	1.79	0.02	0	0
230	**Leg**, average, *raw*, lean and fat	0	0	0	0.03	0.06	0.57	0.07	2.52	0.12	1.96	0.01	0	0
231	**Leg, whole**, *roasted medium*, lean and fat	0	0	0	0.03	0.08	0.10	0.09	2.92	0.15	2.27	0.02	0	0
232	**Loin chops**, *grilled*, lean only	0	0	0	0.03	0.05	0.50	0.07	2.10	0.11	1.77	0.01	0	0
233	**Minced lamb**, *stewed*, lean and fat	0	0	0	0.02	0.06	0.60	0.08	2.57	0.14	2.18	0.02	0	0
234	**Neck fillet**, strips, *stir-fried in corn oil*, lean only	0	0	0	0.03	0.08	0.82	0.10	3.76	0.19	2.82	0.03	0	0
235	**Rack of lamb**, *roasted*, lean and fat	0	0	0	0.08	0.15	1.50	0.22	6.33	0.36	5.52	0.04	0	0
236	**Shoulder**, half bladeside, *pot-roasted*, lean and fat	0	0	0	0.06	0.14	1.35	0.17	5.09	0.29	4.31	0.03	0	0
237	whole, *roasted*, lean and fat	0	0	0	0.06	0.12	1.17	0.16	4.54	0.25	3.67	0.03	0	0
238	**Stewing lamb**, *stewed*, lean and fat	0	0	0	0.05	0.09	0.93	0.13	3.87	0.23	3.48	0.03	0	0

Monounsaturated fatty acids, g per 100g food

No.	Food	10:1	12:1	14:1	cis 15:1	16:1	17:1	18:1	cis/trans 18:1 n-9	18:1 n-7	cis 20:1	22:1	cis/trans 22:1 n-11	22:1 n-9	cis 24:1	trans Monounsatd
	Lamb															
224	**Lamb**, average, trimmed lean, *raw*	0	0	0.01	0.01	0.13	0.05	2.37	2.31	0.43	0.02	0	0	0	0	0.51
225	trimmed fat, *raw*	0	0	0.08	0.02	0.87	0.30	14.47	14.18	3.45	0.14	0	0	0	0	4.22
226	lean only, *cooked*	0	0	0.02	0.01	0.21	0.08	3.71	3.63	0.10	0.03	0	0	0	0	0.80
227	fat only, *cooked*	0	0	0.09	0.03	0.98	0.33	15.97	15.79	3.46	0.14	0	0	0	0	4.46
228	**Best end neck cutlets**, *barbecued*, lean only	0	0	0.02	0.01	0.22	0.08	3.86	3.73	0.80	0.04	0	0	0	0	0.92
229	**Chump steaks/chops**, *fried in corn oil*, lean only	0	0	0.01	0.02	0.21	0.07	3.12	3.01	0.59	0.03	0	0	0	0	0.63
230	**Leg**, average, *raw*, lean and fat	0	0	0.02	0	0.20	0.07	3.73	3.66	0.69	0.02	0	0	0	0	0.82
231	**Leg, whole**, *roasted medium*, lean and fat	0	0	0.03	0.02	0.26	0.10	4.65	4.80	0.56	0.03	0	0	0	0	0.96
232	**Loin chops**, *grilled*, lean only	0	0	0.01	0.01	0.17	0.07	3.02	2.96	0.59	0.02	0	0	0	0	0.74
233	**Minced lamb**, *stewed*, lean and fat	0	0	0.02	0.01	0.24	0.08	3.72	3.59	0.71	0.04	0	0	0	0	0.81
234	**Neck fillet**, strips, *stir-fried in corn oil*, lean only	0	0	0.03	0.01	0.29	0.11	5.90	6.05	0.68	0.09	0	0	0	0	1.15
235	**Rack of lamb**, *roasted*, lean and fat	0	0	0.04	0.01	0.51	0.18	8.44	8.24	1.73	0.08	0	0	0	0	2.06
236	**Shoulder**, half bladeside, *pot-roasted*, lean and fat	0	0	0.04	0.02	0.42	0.16	7.45	7.32	1.47	0.05	0	0	0	0	1.82
237	whole, *roasted*, lean and fat	0	0	0.04	0.03	0.38	0.15	6.54	N	1.22	0.03	0	0	0	0	1.52
238	**Stewing lamb**, *stewed*, lean and fat	0	0	0.03	0.01	0.31	0.12	5.88	5.70	1.16	0.06	0	0	0	0	1.32

Polyunsaturated fatty acids, g per 100g food

No.	Food	cis n-6					cis n-3					trans
		18:2	18:3	20:3	20:4	22:4	18:3	18:4	20:5	22:5	22:6	Polyunsatd
	Lamb											
224	**Lamb**, average, trimmed lean, *raw* [a]	0.13	0.04	0	0.02	0	0.09	0.01	0.02	0.03	0.01	0.09
225	trimmed fat, *raw* [b]	0.60	0.31	0	0.04	0	0.55	0.05	0.02	0.08	0	0.67
226	lean only, *cooked* [a]	0.25	0.07	0	0.03	0.03	0.15	0.01	0.02	0.03	0.01	0.13
227	fat only, *cooked* [c]	0.56	0.35	0	0.03	0	0.61	0.05	0.02	0.07	0	0.73
228	**Best end neck cutlets**, *barbecued*, lean only [d]	0.16	0.08	0	0.03	0	0.14	0.01	0.03	0.04	0.02	0.18
229	**Chump steaks/chops**, *fried in corn oil*, lean only [a]	0.54	0.05	0	0.03	0	0.13	0.01	0.03	0.04	0.01	0.08
230	**Leg**, average, *raw*, lean and fat [e]	0.24	0.06	Tr	0.04	0	0.16	0.01	0.03	0.04	0.02	0.11
231	**Leg, whole**, *roasted medium*, lean and fat [f]	0.22	0.08	0.01	0.04	0	0.17	0.01	0.03	0.04	0.02	0.18
232	**Loin chops**, *grilled*, lean only [g]	0.22	0.05	0	0.02	0	0.13	0.01	0.02	0.03	0.01	0.11
233	**Minced lamb**, *stewed*, lean and fat [a]	0.19	0.07	0	0.03	0	0.13	0.01	0.02	0.04	0	0.08
234	**Neck fillet**, strips, *stir-fried in corn oil*, lean only [h]	1.59	0.10	0	0.03	0	0.22	0.01	0.02	0.04	0	0.15
235	**Rack of lamb**, *roasted*, lean and fat	0.48	0.15	0	0.04	0	0.30	0.02	0.03	0.06	0	0.23
236	**Shoulder**, half bladeside, *pot-roasted*, lean and fat [i]	0.49	0.13	0	0.04	0	0.28	0.02	0.03	0.06	0	0.26
237	whole, *roasted*, lean and fat [e]	0.37	0.11	0	0.04	0	0.22	0.02	0.02	0.05	0	0.19
238	**Stewing lamb**, *stewed*, lean and fat	0.69	0.10	0	0.05	0	0.23	0.01	0.03	0.06	0	0.16

[a] Contains 0.01g 22:2 per 100g food
[b] Contains 0.01g 20:2, 0.04g 22:2 per 100g food
[c] Contains 0.02g 20:2, 0.06g 22:2 per 100g food
[d] Contains 0.01g 20:2, 0.03g 22:2 per 100g food
[e] Contains 0.01g 20:2, 0.02g 22:2 per 100g food

[f] Contains 0.01g 16:4, 0.01g 22:2 per 100g food
[g] Contains 0.01g 20:2, 0.01g 22:2 per 100g food
[h] Contains 0.01g 16:4 per 100g food
[i] Contains 0.02g 20:2, 0.04g 22:2 per 100g food

Fat and total fatty acids, g per 100g food

No.	Food	Description	Total fat	Satd	cis-Mono unsatd	Polyunsatd Total cis	n-6	n-3	Total trans	Total branched
	Pork									
239	**Pork**, average, trimmed lean, *raw*	Average from 8 different cuts	4.0	1.36	1.50	0.69	0.61	0.09	0.02	0.01
240	trimmed fat, *raw*	Average from 5 different cuts	54.8	19.96	22.26	9.34	8.23	1.16	0.34	0.34
241	lean only, *cooked*	Average from 15 different cuts	6.7	2.31	2.56	1.15	1.02	0.12	0.03	0.01
242	fat only, *cooked*	Average from 3 different cuts	48.6	16.42	20.88	8.48	7.40	1.11	0.30	0.08
243	**Belly joint/slices**, *roasted*, lean and fat	10 samples, 65% lean and 35% fat	21.4	7.41	8.39	3.90	3.48	0.45	0.14	0.03
244	**Chump chops/steaks**, *fried in corn oil*, lean and fat	Calculated from 74% lean and 26% fat	11.7	3.78	4.44	2.46	2.21	0.28	0.08	0.02
245	**Diced pork**, *raw*, lean only	10 samples	4.0	1.35	1.58	0.66	0.58	0.08	0.01	0.01
246	*casseroled*, lean only	10 samples	6.4	1.88	2.26	1.56	1.48	0.12	0.02	0.01
247	**Fillet strips**, *stir-fried in corn oil*, lean only	10 samples	5.9	1.32	1.74	2.17	2.13	0.10	0.02	0.01
248	**Leg joint**, *roasted medium*, lean and fat	Calculated from 83% lean and 17% fat	10.2	3.59	4.37	1.42	1.29	0.17	0.04	0.02
249	**Loin chops**, *grilled*, lean and fat	Calculated from 80% lean and 20% fat	15.7	5.53	6.38	2.52	2.22	0.30	0.09	0.03
250	*microwaved*, lean and fat	Calculated from 82% lean and 18% fat	14.1	4.89	5.63	2.45	2.14	0.31	0.07	0.03
251	**Loin joint**, *roasted*, lean and fat	Calculated from 74% lean and 26% fat	16.4	5.89	6.36	2.80	2.48	0.32	0.11	0.02
252	**Loin steaks**, *fried in corn oil*, lean only	22 samples of a mixture of loin steaks, boneless chops and noisettes	7.2	2.31	2.72	1.38	1.28	0.15	0.05	0.01
253	**Spare ribs**, sliced, *grilled*, lean and fat	10 samples	19.5	7.48	8.07	2.34	2.16	0.25	0.12	0.02
254	**Pork steaks**, *grilled*, lean only	19 samples of a mixture of pork and leg steaks	3.7	1.29	1.42	0.58	0.53	0.06	0.02	0.01
255	**Minced pork**, *raw*, lean and fat	10 samples	9.7	3.55	3.82	1.55	1.37	0.19	0.04	0.03
256	*dry fried and stewed*, lean and fat	10 samples	10.4	3.88	4.08	1.56	1.39	0.17	0.08	0.02

No.	Food	4:0	6:0	8:0	10:0	12:0	14:0	15:0	16:0	17:0	18:0	20:0	22:0	24:0
Pork														
239	**Pork**, average, trimmed lean, *raw*	0	0	0	0	0	0.04	Tr	0.83	0.01	0.45	0.01	0	0
240	trimmed fat, *raw*	0	0	0	0.03	0	0.58	0.05	12.19	0.21	6.80	0.10	0	0
241	lean only, *cooked*	0	0	0	0	0.01	0.08	0.01	1.42	0.02	0.76	0.01	0	0
242	fat only, *cooked*	0	0	0	0.03	0.04	0.59	0.04	10.16	0.17	5.30	0.09	0	0
243	**Belly joint/slices**, *roasted*, lean and fat	0	0	0	0.01	0.02	0.27	0.02	4.56	0.09	2.41	0.04	0	0
244	**Chump chops/steaks**, *fried in corn oil,* lean and fat	0	0	0	0.01	0.01	0.13	0.01	2.35	0.04	1.22	0.02	0	0
245	**Diced pork**, *raw*, lean only	0	0	0	Tr	Tr	0.05	Tr	0.84	0.01	0.43	0.01	0	0
246	*casseroled*, lean only	0	0	0	Tr	Tr	0.06	Tr	1.19	0.02	0.60	0.01	0	0
247	**Fillet strips**, *stir-fried in corn oil,* lean only	0	0	0	Tr	Tr	0.03	Tr	0.88	0.01	0.38	0.02	0	0
248	**Leg joint**, *roasted medium*, lean and fat	0	0	0	0.01	0.01	0.11	0.01	2.23	0.04	1.17	0.01	0	0
249	**Loin chops**, *grilled*, lean and fat	0	0	0	0.01	0.01	0.19	0.01	3.42	0.05	1.81	0.03	0	0
250	*microwaved*, lean and fat	0	0	0	0.01	0.01	0.17	0.01	3.04	0.05	1.58	0.02	0	0
251	**Loin joint**, *roasted*, lean and fat	0	0	0	0.01	0.01	0.19	0.02	3.55	0.06	2.02	0.03	0	0
252	**Loin steaks**, *fried in corn oil,* lean only	0	0	0	Tr	0.01	0.07	Tr	1.44	0.02	0.75	0.01	0	0
253	**Spare ribs**, sliced, *grilled*, lean and fat	0	0	0	0.01	0.02	0.25	0.02	4.44	0.06	2.65	0.03	0	0
254	**Pork steaks**, *grilled*, lean only	0	0	0	Tr	Tr	0.04	0.01	0.81	0.01	0.41	0.01	0	0
255	**Minced pork**, *raw*, lean and fat	0	0	0	0.01	0.01	0.13	0.01	2.13	0.04	1.21	0.01	0	0
256	*dry fried and stewed*, lean and fat	0	0	0	0.01	0.01	0.14	0.01	2.34	0.04	1.31	0.02	0	0

Meat, poultry and game

No.	Food	cis 10:1	12:1	14:1	15:1	16:1	17:1	18:1	cis/trans 18:1 n-9	18:1 n-7	cis 20:1	cis 22:1	cis/trans 22:1 n-11	22:1 n-9	cis 24:1	trans Monounsatd
	Pork															
239	**Pork**, average, trimmed lean, *raw*	0	0	0	0	0.09	0.01	1.36	1.26	0.12	0.03	0	0	0	0	0.02
240	trimmed fat, *raw*	0	0	0.16	Tr	1.22	0.17	20.19	18.90	1.48	0.52	0	0	0	0	0.30
241	lean only, *cooked*	0	0	0	0.01	0.15	0.02	2.32	2.14	0.20	0.05	0.01	Tr	Tr	0	0.03
242	fat only, *cooked*	0	0	0.01	0.01	1.18	0.15	19.03	17.79	1.38	0.49	0	0	0	0	0.26
243	**Belly joint/slices**, *roasted*, lean and fat	0	0	0.01	Tr	0.50	0.07	7.61	7.08	0.58	0.20	0	0	0	0	0.10
244	**Chump chops/steaks**, *fried in corn oil*, lean and fat	0	0	0	0.01	0.23	0.04	4.07	3.86	0.35	0.10	0	0	0	0	0.06
245	**Diced pork**, *raw*, lean only	0	0	0	Tr	0.10	0.01	1.44	1.32	0.12	0.03	0	0	0	0	0.01
246	*casseroled*, lean only	0	0	0	0.01	0.12	0.01	2.07	1.92	0.17	0.05	0	0	0	0	0.02
247	**Fillet strips**, *stir-fried in corn oil*, lean only	0	0	0	0.01	0.06	Tr	1.64	1.56	0.09	0.03	0	0	0	0	0.02
248	**Leg joint**, *roasted medium*, lean and fat	0	0	0.01	0.01	0.25	0.02	3.96	3.63	0.34	0.08	0.03	0.01	0.02	0	0.03
249	**Loin chops**, *grilled*, lean and fat	0	0	0.01	0.02	0.38	0.05	5.77	5.34	0.48	0.15	0	0	0	0	0.08
250	*microwaved*, lean and fat	0	0	0.01	0.03	0.33	0.05	5.08	4.72	0.40	0.13	0	0	0	0	0.07
251	**Loin joint**, *roasted*, lean and fat	0	0	Tr	0.01	0.34	0.04	5.81	5.43	0.44	0.16	0	0	0	0	0.11
252	**Loin steaks**, *fried in corn oil*, lean only	0	0	Tr	0.01	0.15	0.01	2.50	2.34	0.19	0.06	0	0	0	0	0.04
253	**Spare ribs**, sliced, *grilled*, lean and fat	0	0	Tr	Tr	0.44	0.05	7.39	6.90	0.56	0.17	0.01	0	0	0	0.11
254	**Pork steaks**, *grilled*, lean only	0	0	Tr	0.01	0.08	0.01	1.28	1.16	0.12	0.03	0.02	0.01	0.01	0	0.01
255	**Minced pork**, *raw*, lean and fat	0	0	0.01	Tr	0.21	0.03	3.48	3.24	0.26	0.08	0	0	0	0	0.03
256	*dry fried and stewed*, lean and fat	0	0	0.01	Tr	0.25	0.03	3.71	3.46	0.30	0.09	0	0	0	0	0.08

Polyunsaturated fatty acids, g per 100g food

No.	Food	cis n-6					18:3	18:4	cis n-3			trans
		18:2	18:3	20:3	20:4	22:4			20:5	22:5	22:6	Polyunsatd
Pork												
239	**Pork**, average, trimmed lean, *raw* [a]	0.54	0	*0.01*	*0.04*	0	0.05	0	0.01	0.02	0.01	Tr
240	trimmed fat, *raw* [b]	7.74	0	*0.16*	*0.09*	0	*0.79*	0	0	0.11	0.11	0.04
241	lean only, *cooked* [c]	0.95	0	*0.02*	*0.04*	0	0.07	0	0.01	0.02	0.01	0.01
242	fat only, *cooked* [d]	6.98	0	0.15	0.08	0	*0.73*	0	0.02	0.12	0.09	0.04
243	**Belly joint/slices**, *roasted*, lean and fat [e]	3.28	0	*0.06*	*0.07*	0	*0.28*	0	0.01	0.05	0.04	0.04
244	**Chump chops/steaks**, *fried in corn oil*, lean and fat[f]	2.04	0	*0.04*	0.08	0	0.17	0	0	0.04	0.03	0.02
245	**Diced pork**, *raw*, lean only [a]	0.52	0	0.01	0.03	0	0.05	0	0.01	0.01	0.01	Tr
246	*casseroled*, lean only [c]	1.38	0	*0.02*	0.06	0	0.07	0	0.01	N	N	Tr
247	**Fillet strips**, *stir-fried in corn oil*, lean only [a]	2.05	0	*0.01*	N	0	0.06	0	0.01	0.02	0.01	Tr
248	**Leg joint**, *roasted medium*, lean and fat [g]	1.18	0	*0.02*	*0.06*	0	*0.08*	0	0.01	0.02	0.01	0.02
249	**Loin chops**, *grilled*, lean and fat [h]	2.04	0	*0.05*	0.06	0	0.19	0	0.01	0.04	0.03	0.01
250	*microwaved*, lean and fat [i]	1.96	0	*0.05*	0.08	0	0.20	0	0.01	0.04	0.03	Tr
251	**Loin joint**, *roasted*, lean and fat [j]	2.29	0	*0.05*	*0.06*	0	0.21	0	0.01	0.04	0.03	Tr
252	**Loin steaks**, *fried in corn oil*, lean only [k]	1.19	0	*0.02*	N	0	0.09	0	0.01	0.02	0.02	0.01
253	**Spare ribs**, sliced, *grilled*, lean and fat [h]	2.00	0	*0.04*	N	0	0.16	0	0.01	0.03	0.02	0.02
254	**Pork steaks**, *grilled*, lean only [a]	0.48	0	0.01	0.03	0	0.03	0	0.00	0.01	0.01	0.01
255	**Minced pork**, *raw*, lean and fat [l]	1.26	0	0.03	0.04	0	0.13	0	0.01	0.02	0.02	0.01
256	*dry fried and stewed*, lean and fat [l]	1.29	0	*0.03*	0.04	0	0.11	0	0	0.02	0.02	0.01

a Contains 0.02g 20:2 per 100g food
b Contains 0.30g 20:2, 0.05g 22:2 per 100g food
c Contains 0.03g 20:2 per 100g food
d Contains 0.28g 20:2, 0.02g 22:2 per 100g food
e Contains 0.12g 20:2 per 100g food
f Contains 0.06g 20:2 per 100g food

g Contains 0.04g 20:2 per 100g food
h Contains 0.08g 20:2, 0.02g 22:2 per 100g food
i Contains 0.07g 20:2, 0.02g 22:2 per 100g food
j Contains 0.01g, 16:4, 0.09g 20:2, 0.02g 22:2 per 100g food
k Contains 0.03g 20:2, 0.01g 22:2 per 100g food
l Contains 0.05g 20:2 per 100g food

Meat, poultry and game

Fat and total fatty acids, g per 100g food

No.	Food	Description	Total fat	Satd	cis-Mono unsatd	Polyunsatd Total cis	n-6	n-3	Total trans	Total branched
Poultry										
257	**Chicken**, dark meat, *raw*	31 samples	2.8	0.74	1.28	0.55	0.46	0.09	0.02	Tr
258	light meat, *raw*	31 samples	1.1	0.31	0.48	0.22	0.18	0.04	0.01	Tr
259	skin, *raw*	25 samples	48.3	13.40	23.06	7.89	6.71	1.35	0.57	0.17
260	-, *cooked*	30 samples of roasted and grilled skin, dry	46.1	12.78	22.00	7.26	6.40	1.29	0.54	0.56
261	dark meat, *roasted*	19 samples of fresh and frozen chicken	10.9	2.89	5.00	2.13	1.81	0.33	0.08	0.04
262	light meat, *roasted*	19 samples of fresh and frozen chicken	3.7	1.02	1.58	0.72	0.60	0.13	0.05	0.02
263	breast, *grilled*, meat only	27 samples	2.7	0.75	1.13	0.52	0.41	0.11	0.03	0.01
264	portions, *casseroled* with skin, dark meat	10 samples	9.7	2.58	4.37	1.90	1.58	0.34	0.13	0.03
265	-,-, light meat	10 samples	5.1	1.42	2.29	0.93	0.77	0.16	0.06	0.02
266	-, *fried*, meat only, takeaway	10 samples from different outlets	9.7	2.64	4.44	1.69	1.43	0.30	0.19	0.03
267	corn-fed, *raw*, dark meat	6 samples	7.2	2.06	3.14	1.46	1.32	0.13	0.03	0.02
268	-,-, light meat	6 samples	2.8	0.81	1.19	0.57	0.51	0.06	0.01	0.01
269	**Turkey**, skin, *raw*	21 samples	30.7	9.97	11.51	6.64	5.83	0.90	0.50	0.15
270	dark meat, *roasted*	27 samples including self basting turkey	7.0	2.10	2.48	1.74	1.50	0.27	0.11	0.03
271	light meat, *roasted*	18 samples	1.9	0.62	0.67	0.43	0.37	0.06	0.01	0.01
272	breast fillet, *grilled*, meat only	9 samples, skinless	1.7	0.61	0.56	0.35	0.31	0.04	0.02	0.01
273	minced, *stewed*	5 samples	6.8	1.95	2.18	2.11	1.92	0.20	0.07	0.03
274	thighs, diced, *casseroled*	8 samples, skinless	7.5	2.45	2.59	1.79	1.63	0.16	0.08	0.03

Meat, poultry and game

Saturated fatty acids, g per 100g food

Poultry

No.	Food	4:0	6:0	8:0	10:0	12:0	14:0	15:0	16:0	17:0	18:0	20:0	22:0	24:0
257	**Chicken**, dark meat, *raw*	0	0	0	0	0	0.02	Tr	0.54	0.01	0.16	0	0	0
258	light meat, *raw*	0	0	0	0	0	0.01	Tr	0.22	Tr	0.07	0	0	0
259	skin, *raw*	0	0	0	0	0.08	0.45	0.07	9.83	0.12	2.78	0.05	0.01	0
260	-, *cooked*	0	0	0	0	0.07	0.43	0.07	9.38	0.12	2.65	0.05	0.01	0
261	dark meat, *roasted*	0	0	0	0	0.02	0.09	0.01	2.11	0.02	0.62	0.01	0	0
262	light meat, *roasted*	0	0	0	0	Tr	0.03	Tr	0.74	0.01	0.22	Tr	Tr	0
263	breast, *grilled*, meat only	0	0	0	0	Tr	0.02	0	0.54	0.01	0.17	Tr	0	0
264	portions, *casseroled* with skin, dark meat	0	0	Tr	0	0.01	0.07	0.01	1.88	0.02	0.56	0.01	0	0
265	-, light meat	0	0	0.01	Tr	0	0.04	0.01	1.04	0.01	0.31	Tr	0	0
266	-, *fried*, meat only, takeaway	0	0	0.01	Tr	0.01	0.08	0.01	1.91	0.02	0.56	0.02	0.02	0.01
267	corn-fed, *raw*, dark meat	0	0	Tr	Tr	0.01	0.05	0.01	1.54	0.01	0.44	Tr	0	0
268	-, light meat	0	0	Tr	Tr	Tr	0.02	Tr	0.61	0.01	0.17	Tr	0	0
269	**Turkey**, skin, *raw*	0	0	Tr	0	0.03	0.43	0.08	7.00	0.11	2.27	0.04	0.01	0
270	dark meat, *roasted*	0	0	0	0	0.01	0.09	0.02	1.36	0.03	0.59	0.01	0	0
271	light meat, *roasted*	0	0	Tr	Tr	Tr	0.02	Tr	0.40	0.01	0.19	Tr	0	0
272	breast fillet, *grilled*, meat only	0	0	0	0	Tr	0.02	Tr	0.41	0.01	0.17	Tr	0	0
273	minced, *stewed*	0	0	Tr	Tr	0.01	0.06	0.01	1.31	0.02	0.52	0.01	0	0
274	thighs, diced, *casseroled*	0	0	Tr	Tr	0.01	0.08	0.01	1.65	0.02	0.66	0.01	0	0

Poultry

No.	Food	cis 10:1	12:1	14:1	15:1	16:1	17:1	18:1	cis/trans 18:1 n-9	18:1 n-7	cis 20:1	cis 22:1	cis/trans 22:1 n-11	22:1 n-9	cis 24:1	trans Monounsatd
257	**Chicken**, dark meat, *raw*	0	0	0	0	0.13	0.01	1.13	1.07	0.07	0.02	Tr	0	0	0	0.02
258	light meat, *raw*	0	0	0	0	0.04	Tr	0.42	0.40	0.03	0.01	Tr	0	0	0	0.01
259	skin, *raw*	0	0	0.09	0	2.52	0.10	20.10	19.24	1.17	0.24	0.02	0	0.02	0	0.51
260	-, *cooked*	0	0	0.08	0	2.40	0.10	19.16	18.35	1.12	0.23	0.02	0	0.02	0	0.49
261	dark meat, *roasted*	0	0	0.02	0	0.49	0.03	4.39	4.16	0.27	0.07	0.01	0	0.01	0	0.07
262	light meat, *roasted*	0	0	Tr	0	0.14	0.01	1.41	1.33	0.10	0.02	Tr	0	0	0	0.04
263	breast, *grilled*, meat only	0	0	Tr	0	0.11	Tr	1.00	0.95	0.06	0.01	Tr	0	0	0	0.02
264	portions, *casseroled* with skin, dark meat	0	0	0.01	0	0.44	0.01	3.85	3.68	0.24	0.05	Tr	0	0	0	0.10
265	-,- light meat	0	0	0.01	0	0.23	0.01	2.02	1.92	0.14	0.03	Tr	0	0	0	0.04
266	-, *fried*, meat only, takeaway	0	0	0.01	0	0.38	Tr	4.00	3.89	0.21	0.05	Tr	0	0	0	0.16
267	corn-fed, *raw*, dark meat	0	0	0.01	0	0.37	0.01	2.72	2.59	0.15	0.03	Tr	0	0	0	0.03
268	-,- light meat	0	0	0	0	0.13	Tr	1.04	1.00	0.04	0.01	Tr	0	0	0	0.01
269	**Turkey**, skin, *raw*	0	0	0.07	0	1.56	0.09	9.56	9.14	0.69	0.17	0.06	0.03	0.03	0	0.40
270	dark meat, *roasted*	0	0	0.01	0	0.24	0.02	2.15	2.08	0.14	0.05	0.01	0.01	0.01	0	0.09
271	light meat, *roasted*	0	0	0	0	0.06	Tr	0.58	0.53	0.05	0.01	0.01	0	0.01	0	0.01
272	breast fillet, *grilled*, meat only	0	0	0	0	0.07	Tr	0.47	0.43	0.04	0.01	0.01	0	0.01	0	0.01
273	minced, *stewed*	0	0	0.01	0	0.22	0.01	1.90	1.84	0.10	0.03	0.01	0	0.01	0	0.05
274	thighs, diced, *casseroled*	0	0	0.01	0	0.34	0.02	2.19	2.11	0.14	0.03	0.01	0	0.01	0	0.07

Meat, poultry and game

Polyunsaturated fatty acids, g per 100g food

No.	Food	cis n-6					cis n-3					trans
		18:2	18:3	20:3	20:4	22:4	18:3	18:4	20:5	22:5	22:6	Polyunsatd
Poultry												
257	**Chicken**, dark meat, *raw* a	0.44	0	Tr	0.01	0	0.07	0	0	0.01	0.01	Tr
258	light meat, *raw*	0.16	0	Tr	0.01	0	0.02	0	Tr	0.01	0.01	Tr
259	skin, *raw* b	6.54	0	0.01	0.01	0	1.07	0	0.09	0.05	0.06	0.16
260	–, *cooked*	6.23	0	Tr	Tr	0	1.02	0	0	0	0	0.15
261	dark meat, *roasted* c	1.72	0	0.01	0.06	0	0.27	0	0	0.03	0.03	0.01
262	light meat, *roasted* a	0.55	0	0.01	0.03	0	0.07	0	0.01	0.02	0.03	0.01
263	breast, *grilled,* meat only d	0.38	0	0.01	0.02	0	0.05	0	0.01	0.02	0.03	0.01
264	portions, *casseroled* with skin, dark meat d	1.49	0	0.02	0.05	0	0.22	0	0.03	0.03	0.04	0.03
265	–, –, light meat d	0.73	0	0.01	0.03	0	0.10	0	0.01	0.02	0.03	0.01
266	–, *fried,* meat only, takeaway e	1.33	0	0.01	0.04	0	0.19	0.01	0.02	0.03	0.03	0.04
267	corn-fed, *raw,* dark meat d	0.46	0	0.01	0.03	0	0.03	0	0	0.01	0.01	Tr
268	–, –, light meat d	0.46	0	0.01	0.03	0	0.03	0	0	0.01	0.01	Tr
269	**Turkey**, skin, *raw* f	5.65	0	0.03	0.08	0	0.57	0.02	0.09	0.06	0.10	0.09
270	dark meat, *roasted* a	1.38	0	0.01	0.08	0	0.15	0	0.02	0.02	0.05	0.03
271	light meat, *roasted*	0.33	0	0.01	0.03	0	0.03	0	0.01	0.01	0.02	Tr
272	breast fillet, *grilled,* meat only	0.27	0	Tr	0.03	0	0.01	0	0	0.01	0.01	Tr
273	minced, *stewed* a	1.81	0	0.01	0.08	0	0.12	0	0.02	0.02	0.04	0.02
274	thighs, diced, *casseroled* a	1.53	0	0.01	0.08	0	0.11	0	0.01	0.02	0.02	0.01

a Contains 0.01g 20:2 per 100g food
b Contains 0.06g 20:2 per 100g food
c Contains 0.02g 20:2 per 100g food

d Contains 0.01g 16:4, 0.01g 20:2 per 100g food
e Contains 0.01g 16:4, 0.02g 20:2 per 100g food
f Contains 0.03g 20:2 per 100g food

Meat, poultry and game

Fat and total fatty acids, g per 100g food

No.	Food	Description	Total fat	Satd	cis-Mono unsatd	Polyunsatd Total cis	n-6	n-3	Total trans	Total branched
Other poultry										
275	**Duckling**, *raw*, meat only	14 samples, meat from dressed carcase	6.5	2.01	3.11	0.96	0.88	0.09	0.07	0
276	-, meat, fat and skin	14 samples, meat from dressed carcase	42.7	12.23	21.39	6.38	5.73	0.69	0.40	0
277	*roasted*, meat only	14 samples, meat from dressed carcase	10.4	3.29	5.10	1.31	1.23	0.11	0.14	0
278	-, meat, fat and skin	14 samples, meat from dressed carcase	49.6	14.91	24.70	6.75	6.14	0.70	0.52	0
Game										
279	**Pheasant**, *roasted*, meat only	3 samples	3.2	1.08	1.48	0.42	0.40	0.02	0.02	0.02
280	*casseroled*, meat only	3 samples	15.6	5.35	7.34	1.92	1.81	0.12	0.09	0.04
281	**Rabbit**, *stewed*, meat and fat	4 samples	3.4	1.79	0.67	0.64	0.41	0.25	0.06	0.05
282	**Venison**, *raw*	8 samples including diced and steaks	2.2	0.94	0.45	0.48	0.32	0.17	0.08	0.09
283	*casseroled*, meat only	8 samples including diced and steaks	3.7	1.62	0.78	0.74	0.52	0.26	0.15	0.17
Offal										
284	**Heart**, lamb, *raw*	10 samples from different outlets	6.8	2.84	1.57	0.49	0.46	0.13	0.46	0.05
285	pig, *raw*	6 samples from different outlets	3.8	1.41	1.03	0.51	0.49	0.02	0.03	0.03
286	**Kidney**, lamb, grilled	10 samples from different outlets	4.6	1.35	0.92	0.98	0.85	0.17	0.18	0.02
287	pig, *raw*	Data from Institute of Human Nutrition and Brain Chemistry	2.7	0.90	0.71	0.41	0.40	0.01	0	0
288	**Liver**, chicken, *raw*	Data from Institute of Human Nutrition and Brain Chemistry	2.3	0.72	0.51	0.47	0.36	0.11	0	0

Saturated fatty acids, g per 100g food

No.	Food	4:0	6:0	8:0	10:0	12:0	14:0	15:0	16:0	17:0	18:0	20:0	22:0	24:0
Other poultry														
275	**Duck**, *raw*, meat only	0	0	0.01	0	0	0.04	0	1.37	0.02	0.52	0.02	0.02	0.02
276	-, meat, fat and skin	0	0	0.04	0	0.04	N	0	9.16	0.12	2.62	0.12	0.08	0.04
277	*roasted*, meat only	0	0	Tr	0	0.01	0.06	0	2.31	0.03	0.82	0.03	0.02	0.02
278	-, meat, fat and skin	0	0	0.05	0	0.05	0.33	0	10.92	0.14	3.19	0.09	0.09	0.05
Game														
279	**Pheasant**, *roasted*, meat only	0	0	0	0	0	0.02	Tr	0.71	0.01	0.32	0.01	0.01	0
280	*casseroled*, meat only	0	0	0	0	0	0.12	0.06	3.95	0.01	1.16	0.01	0.01	0
281	**Rabbit**, *stewed*, meat and fat	0	0.01	0.01	0	0	0.07	0.02	1.11	0.04	0.51	0.01	0.01	0.01
282	**Venison**, *raw*	0	0	0	0	0	0.10	0.04	0.41	0.02	0.35	Tr	0.01	0
283	*casseroled*, meat only	0	0	0	0	0.01	0.18	0.06	0.71	0.04	0.60	0.01	0.01	0
Offal														
284	**Heart**, lamb, *raw*	0	0	0	0.01	0.02	0.16	0.03	0.97	0.10	1.53	0.01	0.01	0.01
285	pig, *raw*	0	0	0	0	0	0.04	0	0.73	0.01	0.59	0.01	0.01	0.01
286	**Kidney**, lamb, *grilled*	0	0	0	0	0	0.02	0.01	0.59	0.03	0.58	0.01	0.05	0.05
287	pig, *raw*	0	0	0	0	0	0.02	0	0.51	0	0.37	0	0	0
288	**Liver**, chicken, *raw*	0	0	0	0	0	0.01	0	0.43	0	0.29	0	0	0

Meat, poultry and game

Monounsaturated fatty acids, g per 100g food

No.	Food	cis 10:1	12:1	14:1	15:1	16:1	17:1	18:1	cis/trans 18:1 n-9	18:1 n-7	cis 20:1	22:1	cis/trans 22:1 n-11	22:1 n-9	cis 24:1	trans Monounsatd
Other poultry																
275	**Duck**, *raw*, meat only	0	0	0	0	0.18	0.04	2.83	2.61	0	0.06	0.01	Tr	Tr	0	0.06
276	-, meat, fat and skin	0	0	0	0	1.41	0.08	19.57	18.32	0	0.32	0	0	0	0	0.36
277	*roasted*, meat only	0	0	0	0	0.29	0.02	4.72	4.37	0	0.06	0.01	Tr	Tr	0	0.11
278	-, meat, fat and skin	0	0	0	0	1.50	0.09	22.73	21.28	1.45	0.33	0.05	N	N	0	0.42
Game																
279	**Pheasant**, *roasted*, meat only	0	0	0	0	0.15	Tr	1.23	1.07	0	0.01	0.09	N	N	0	0.02
280	*casseroled*, meat only	0	0	0	0	1.03	0.06	6.15	5.54	0	0.04	0.06	N	N	0	0.07
281	**Rabbit**, *stewed*, meat and fat	0	0	0	0	0.04	0.04	0.56	0.51	0	0.01	0.02	N	N	0	0.04
282	**Venison**, *raw*	0	0	0	0	0.06	0.02	0.36	0.31	0	0	0	0	0	0	0.06
283	*casseroled*, meat only	0	0	0	0	0.12	0.04	0.61	0.52	0	0.01	0	0	0	0	0.11
Offal																
284	**Heart**, lamb, *raw*	0	0	0	0	0.04	0.04	1.45	0	1.33	0.04	0	0	0	0.01	0.36
285	pig, *raw*	0	0	0	0	0.04	0.02	0.94	0.82	0	0.02	0	0	0	0.01	0.03
286	**Kidney**, lamb, *grilled*	0	0	0	0	0.06	0.03	0.78	0.71	0	0.02	0	0	0	0.02	0.13
287	pig, *raw*	0	0	0	0	0.04	0	0.66	0.66	0	0.01	0	0	0	0	0
288	**Liver**, chicken, *raw*	0	0	0	0	0.06	0	0.45	0.45	0	0	0	0	0	0	0

Meat, poultry and game

Polyunsaturated fatty acids, g per 100g food

No.	Food	cis n-6					cis n-3					trans
		18:2	18:3	20:3	20:4	22:4	18:3	18:4	20:5	22:5	22:6	Polyunsatd
Other poultry												
275	**Duck**, *raw*, meat only	0.87	0	0	0	0	0.09	0	0	0	0	0.01
276	-, meat, fat and skin	5.69	0	0	0	0	0.69	0	0	0	0	0.04
277	*roasted*, meat only	1.20	0	0	0	0	0.11	0	0	0	0	0.03
278	-, meat, fat and skin	6.05	0	0	0	0	0.70	0	0	0	0	0.09
Game												
279	**Pheasant**, *roasted*, meat only	0.40	0	0	0	0	0.02	0	0	0	0	0
280	*casseroled*, meat only	1.80	0	0	0	0	0.12	0	0	0	0	0.01
281	**Rabbit**, *stewed*, meat and fat	0.39	0	0	0	0	0.25	0	0	0	0	0.02
282	**Venison**, *raw*	0.22	0	0	0.09	0	0.08	0	0.03	0.07	0	0.02
283	*casseroled*, meat only	0.34	0	0	0.14	0	0.12	0	0.04	0.09	0	0.03
Offal												
284	**Heart**, lamb, *raw*	0.31	0	0	0.05	0	0.09	0	0.04	0	0	0.09
285	pig, *raw*	0.42	0	0	0.07	0	0.01	0	0.01	0	0	Tr
286	**Kidney**, lamb, *grilled*	0.64	0	0	0.19	0	0.05	0	0.09	0	0	0.05
287	pig, *raw*	0.24	0	0.01	0.14	0	0.01	0	0	0	0	0
288	**Liver**, chicken, *raw*	0.26	0	0.01	0.10	0	0.01	0	0	0.01	0.08	0

Meat products and dishes

Fat and total fatty acids, g per 100g food

No.	Food	Description	Total fat	Satd	cis-Mono unsatd	Polyunsatd Total cis	n-6	n-3	Total trans	Total branched
Offal *continued*										
289	**Liver**, lamb, *raw*	Data from Institute of Human Nutrition and Brain Chemistry	6.2	1.75	1.87	0.97	0.42	0.55	0	0
290	pig, *raw*	Data from Institute of Human Nutrition and Brain Chemistry	3.1	0.95	0.49	0.85	0.71	0.15	0	0
291	**Tripe**, dressed, *stewed*	6 samples from different outlets	0.5	0.19	0.16	0.02	0.02	0.01	0.02	0.01
Bacon and ham										
292	**Bacon rashers, back**, *raw*	10 samples; smoked and unsmoked, loose and prepacked British, Danish and Dutch bacon	16.5	6.16	6.86	2.24	2.00	0.24	0.08	0.02
293	*grilled*	15 samples; smoked and unsmoked, loose and prepacked British, Danish and Dutch bacon	21.6	7.91	8.85	2.71	2.41	0.31	0.01	0.04
294	**middle**, *raw*	9 samples; smoked and unsmoked, loose and prepacked British, Danish and Dutch bacon	20.0	7.14	8.23	2.27	2.23	0.26	0.14	0.04
295	–, *fried in corn oil*	9 samples; smoked and unsmoked, loose and prepacked British and Danish bacon	28.5	9.96	12.03	4.48	4.40	0.53	0.13	0.05
296	–, *grilled*	9 samples; smoked and unsmoked, loose and prepacked British and Danish bacon	23.1	8.15	9.55	2.96	2.59	0.38	0.14	0.04
297	**streaky**, *raw*	10 samples; smoked and unsmoked, loose and prepacked British and Danish bacon	23.6	7.97	9.82	3.07	3.04	0.37	0.10	0.04
298	–, *grilled*	10 samples; smoked and unsmoked, loose and prepacked British and Danish bacon	26.9	9.53	11.02	3.17	3.22	0.39	0.10	0.04

Meat products and dishes

Saturated fatty acids, g per 100g food

No.	Food	4:0	6:0	8:0	10:0	12:0	14:0	15:0	16:0	17:0	18:0	20:0	22:0	24:0
Offal *continued*
289	**Liver**, lamb, *raw*	0	0	0	0	0.01	0.07	0	0.87	0	0.69	0.09	0.01	0
290	pig, *raw*	0	0	0	0	0	0	0	0.41	0	0.54	0	0	0
291	**Tripe**, dressed, *stewed*	0	0	0	0	0	0.01	0	0.09	0.01	0.08	0	0	0
Bacon and ham
292	**Bacon rashers, back**, *raw*	0	0	0	0.01	0.02	0.22	0.01	3.76	0.05	2.08	0.03	0	0
293	*grilled*	0	0	0	0.01	0.02	0.28	0.01	4.77	0.06	2.71	0.04	0	0
294	**middle**, *raw*	0	0	0	0.01	0.04	0.30	0.01	4.32	0.06	2.36	0.03	0	0
295	-, *fried in corn oil*	0	0	0	0.02	0.04	0.40	0.02	6.10	0.08	3.25	0.05	0	0
296	-, *grilled*	0	0	0	0.01	0.04	0.33	0.01	4.92	0.07	2.72	0.04	0	0
297	**streaky**, *raw*	0	0	0	0.02	0.03	0.31	0.02	5.02	0.07	2.48	0.04	0	0
298	-, *grilled*	0	0	0	0.02	0.02	0.35	0.02	5.95	0.08	3.04	0.05	0	0

Meat products and dishes

Monounsaturated fatty acids, g per 100g food

No.	Food	10:1	12:1	14:1	cis 15:1	16:1	17:1	18:1	cis/trans 18:1 n-9	18:1 n-7	20:1	cis 22:1	cis/trans 22:1 n-11	22:1 n-9	cis 24:1	trans Monounsatd
Offal continued																
289	**Liver**, lamb, *raw*	0	0	0	0	0.17	0	1.70	1.70	0	0.01	0	0	0	0	0
290	pig, *raw*	0	0	0	0	0.04	0	0.45	0.45	0	0	0	0	0	0	0
291	**Tripe**, dressed, *stewed*	0	0	0	0	0.01	Tr	0.14	0.14	0.01	0	0	0	0	0	0.01
Bacon and ham																
292	**Bacon rashers**, back, *raw*	0	0	0	0	0.38	0.04	6.31	5.90	0.45	0.13	0	0	0	0	0.07
293	*grilled*	0	0	0	0	0.48	0.05	8.13	7.51	0.59	0.17	0	0	0	0	N
294	**middle**, *raw*	0	0	0	0	0.48	0	7.57	7.04	0.59	0.18	0	0	0	0	0.12
295	-, *fried in corn oil*	0	0	0	0	0.66	0	11.10	10.26	0.83	0.27	0	0	0	0	0.12
296	-, *grilled*	0	0	0	0	0.53	0	8.79	8.17	0.69	0.22	0	0	0	0	0.13
297	**streaky**, *raw*	0	0	0	0	0.65	0	9.02	8.35	0.71	0.15	0	0	0	0	0.10
298	-, *grilled*	0	0	0	0	0.72	0	10.08	9.30	0.81	0.23	0	0	0	0	0.09

Meat, poultry and game

Polyunsaturated fatty acids, g per 100g food

No.	Food	cis n-6					18:3	18:4	cis n-3			trans Polyunsatd
		18:2	18:3	20:3	20:4	22:4			20:5	22:5	22:6	
Offal continued												
289	**Liver**, lamb, *raw* [a]	0.23	0	0.02	0.09	0.01	0.17	0	0.10	*0.16*	0.13	0
290	pig, *raw*	0.34	0	0.03	0.33	0	0.01	0	0	0.06	0.08	0
291	**Tripe**, dressed, *stewed*	0.01	0	0	0	0	0	0	0	0	0	0
Bacon and ham												
292	**Bacon rashers, back**, *raw* [b]	1.88	0	*0.04*	0.04	0	0.19	0	0	0.03	0	0.01
293	*grilled* [c]	2.22	0	*0.05*	0.08	0	0.22	0	0.01	0.03	0.02	0.01
294	**middle**, *raw* [b]	2.08	0	*0.05*	0.04	0	0.20	0	0	0.03	0	0.02
295	*-, fried in corn oil* [d]	4.18	0	*0.06*	0.10	0	0.30	0	0.03	0.06	0.08	0.01
296	*-, grilled* [e]	2.38	0	*0.06*	0.09	0	0.26	0	0.01	0.04	0.03	0.01
297	**streaky**, *raw* [e]	2.79	0	*0.06*	0.08	0	0.29	0	0.02	*0.08*	0	0.01
298	*-, grilled* [f]	2.91	0	*0.06*	0.08	0	0.27	0	0.01	0.04	0.03	0.01

a Contains 0.07g 20:2 per 100g food
b Contains 0.08g 20:2 per 100g food
c Contains 0.09g 20:2 per 100g food

d Contains 0.12g 20:2 per 100g food
e Contains 0.10g 20:2 per 100g food
f Contains 0.12g 20:2, 0.09g 22:2 per 100g food

Meat products and dishes

Fat and total fatty acids, g per 100g food

No.	Food	Description	Total fat	Satd	cis-Mono unsatd	Polyunsatd Total cis	n-6	n-3	Total trans	Total branched
	Bacon and ham continued									
299	**Bacon loin steaks**, *grilled*	7 samples; smoked and unsmoked, loose and prepacked Danish bacon	9.7	3.43	3.95	1.32	1.17	0.15	0.04	0.01
300	**Ham, gammon joint**, *boiled*	10 samples; smoked and unsmoked, loose and prepacked British and Danish gammon	12.3	4.11	5.35	1.93	1.68	0.26	0.03	0.01
301	**gammon rashers**, *grilled*	5 samples; unsmoked British gammon	9.9	3.42	4.01	1.65	1.23	0.19	0.06	0.06
302	**Ham**, regular	10 samples, 9 brands; loose and prepacked including honey roasted and smoked ham. Added water 10-15%	3.3	1.08	1.37	0.51	0.44	0.06	0.03	0.02
303	canned	10 samples, 6 brands	4.5	1.64	2.05	*0.44*	*0.40*	N	N	N
304	premium	10 samples, 7 brands; loose and prepacked including honey glazed and smoked ham. No added water	5.0	1.64	2.07	0.77	0.67	0.09	0.04	0.02
305	**Pork shoulder**, cured, slices	8 samples, 7 brands	3.6	1.18	1.49	0.55	0.48	0.07	0.03	0.02
	Beefburgers and grillsteaks									
306	**Beefburgers**, chilled/frozen, *raw*	8 samples, 3 brands. 98-99% meat	24.7	10.71	9.97	0.44	0.37	0.12	1.43	1.43
307	*fried in vegetable oil*	8 samples, 3 brands	23.9	10.27	10.12	0.42	0.38	0.11	0.78	0.45
308	reduced fat, chilled/frozen, *raw*	11 samples, 4 brands	9.5	4.19	3.95	0.21	0.18	0.05	0.24	0.19
309	-, *fried in vegetable oil*	11 samples, 4 brands	10.8	4.73	4.39	0.37	0.34	0.07	0.31	0.23
310	**Economy burgers**, *grilled*	10 samples, 6 brands containing onion	19.3	7.06	8.08	1.75	1.67	0.25	0.69	0.20
311	**Grillsteaks**, beef, *raw*	10 samples, 7 brands. 95-98% meat	22.2	9.52	9.38	0.44	0.33	0.10	0.60	0.48

No.	Food	4:0	6:0	8:0	10:0	12:0	14:0	15:0	16:0	17:0	18:0	20:0	22:0	24:0
Bacon and ham continued														
299	**Bacon loin steaks**, *grilled*	0	0	0	0.01	0.01	0.13	0.01	2.10	0.03	1.12	0.02	0	0
300	**Ham, gammon joint**, *boiled*	0	0	0	0.01	0.01	0.15	0.01	2.61	0.03	1.28	0.02	0	0
301	**gammon rashers**, *grilled*	0	0	0	0.01	0.01	0.12	0.01	2.07	0.03	1.17	0.02	0	0
302	**Ham**, regular	0	0	0	0	0	0.04	0	0.68	0.01	0.34	0	0	0
303	canned	0	0	0	0	0	0.06	0	1.09	Tr	0.49	0	0	0
304	premium	0	0	0	0	0	0.06	0	1.03	0.01	0.52	0.01	0	0
305	**Pork shoulder**, cured, slices	0	0	0	0	0	0.04	0	0.74	0.01	0.37	0	0	0
Beefburgers and grillsteaks														
306	**Beefburgers**, chilled/frozen, *raw*	0	0	0	0	0.02	0.76	0.12	6.05	0.25	3.49	0.02	0	0
307	*fried in vegetable oil*	0	0	0	0	0.02	0.71	0.11	5.76	0.25	3.39	0.02	0	0
308	reduced fat, chilled/frozen, *raw*	0	0	0	0	0.01	0.28	0.05	2.33	0.09	1.42	0.01	0	0
309	-, *fried in vegetable oil*	0	0	0.01	0.01	0.01	0.33	0.05	2.64	0.10	1.57	0.01	0	0
310	**Economy burgers**, *grilled*	0	0	0.01	0	0.02	0.39	0.06	4.35	0.13	2.08	0.03	0	0
311	**Grillsteaks**, beef, *raw*	0	0	0	0	0.02	0.66	0.10	5.35	0.23	3.13	0.02	0	0

Meat products and dishes

Monounsaturated fatty acids, g per 100g food

No.	Food	cis 10:1	12:1	14:1	15:1	16:1	17:1	18:1	cis/trans 18:1 n-9	18:1 n-7	cis 20:1	22:1	cis/trans 22:1 n-11	22:1 n-9	cis 24:1	trans Monounsatd
	Bacon and ham continued															
299	**Bacon loin steaks**, grilled	0	0	0	0	0.24	0.02	3.60	3.33	0.29	0.09	0	0	0	0	0.04
300	**Ham, gammon joint**, boiled	0	0	0	0	0.35	0.04	4.85	4.45	0.41	0.11	0	0	0	0	0.02
301	**gammon rashers**, grilled	0	0	0	0	0.21	0.03	3.69	3.43	0.29	0.08	0	0	0	0	0.06
302	**Ham**, regular	0	0	0	0	0.09	0.01	1.25	1.15	0.11	0.03	0	0	0	0	0.02
303	canned	0	0	0	0	0.16	0	1.86	N	N	0.03	0	0	0	0	N
304	premium	0	0	0	0	0.13	0.01	1.89	1.74	0.17	0.04	0	0	0	0	0.03
305	**Pork shoulder**, cured, slices	0	0	0	0	0.09	0.01	1.36	1.25	0.12	0.03	0	0	0	0	0.02
	Beefburgers and grillsteaks															
306	**Beefburgers**, chilled/frozen, raw	0	0	0.23	0	0.99	0.21	8.40	8.08	1.34	0.14	0	0	0	0	1.38
307	fried in vegetable oil	0	0	0.22	0	0.98	0.20	8.64	8.29	0.76	0.07	0	0	0	0	0.71
308	reduced fat, chilled/frozen, raw	0	0	0.08	0	0.36	0.08	3.39	3.19	0.30	0.04	0	0	0	0	0.22
309	-, fried in vegetable oil	0	0	0.09	0.02	0.41	0.09	3.75	3.56	0.33	0.02	0	0	0	0	0.27
310	**Economy burgers**, grilled	0	0	0.09	0	0.73	0.10	7.02	6.93	0.43	0.13	0.02	0	0.02	0	0.51
311	**Grillsteaks**, beef, raw	0	0	0.19	0	0.85	0.19	8.11	7.70	0.75	0.04	0	0	0	0	0.60

Meat products and dishes

Polyunsaturated fatty acids, g per 100g food

No.	Food	cis n-6					cis n-3					trans
		18:2	18:3	20:3	20:4	22:4	18:3	18:4	20:5	22:5	22:6	Polyunsatd
299	**Bacon loin steaks**, *grilled* [a]	1.04	0	0.02	0.05	0	0.09	0	0.01	0.02	0.02	0
300	**Ham, gammon joint**, *boiled* [b]	1.50	0	0.04	0.06	0	0.16	0	0.01	0.04	0.03	0.01
301	**gammon rashers**, *grilled* [c]	1.09	0	0.02	0.06	0	0.13	0	0.01	0.02	0.01	0
302	**Ham**, regular [d]	0.40	0	0.01	0.04	0	0.03	0	0	0.01	0.01	0.01
303	canned	*0.40*	*0*	*0*	*0.04*	*0*	*0*	*0*	*0*	*0*	*0*	*Tr*
304	premium [d]	0.61	0	0.02	0.05	0	0.05	0	0.01	0.02	0.01	0.01
305	**Pork shoulder**, cured, slices [d]	0.44	0	0.01	0.04	0	0.04	0	0.01	0.01	0.01	0.01
	Beefburgers and grillsteaks											
306	**Beefburgers**, chilled/frozen, *raw*	*0.31*	*0*	*0*	*0.02*	*0*	*0.12*	*0*	*0*	*0*	*0*	*0.05*
307	*fried in vegetable oil*	*0.27*	*0*	*0*	*0.04*	*0*	*0.11*	*0*	*0*	*0*	*0*	*0.07*
308	reduced fat, chilled/frozen, *raw*	0.15	0	0	0.01	0	0.05	0	0	0	0	0.02
309	*-, fried in vegetable oil*	*0.28*	*0*	*0*	*0.02*	*0*	*0.07*	*0*	*0*	*0*	*0*	*0.04*
310	**Economy burgers**, *grilled* [e]	*1.45*	*0*	*0.02*	*0.04*	*0*	*0.18*	*0*	*0.01*	*0.01*	*0.01*	*0.18*
311	**Grillsteaks**, beef, *raw*	0.33	0	0	0	0	0.10	0	0	0.01	0	0

[a] Contains 0.04g 20:2, 0.03g 22:2 per 100g food
[b] Contains 0.06g 20:2, 0.04g 22:2 per 100g food
[c] Contains 0.06g 20:2, 0.03g 22:2 per 100g food
[d] Contains 0.01g 16:4 per 100g food
[e] Contains 0.03g 20:2 per 100g food

Meat products and dishes

Fat and total fatty acids, g per 100g food

No.	Food	Description	Total fat	Satd	cis-Mono unsatd	Polyunsatd Total cis	n-6	n-3	Total trans	Total branched
Meat pies and pastries										
312	**Beef pie**, chilled/frozen, *baked*	20 samples, 6 brands	19.4	8.50	6.77	1.87	1.71	0.29	1.22	0.10
313	**Chicken pie**, individual, chilled/frozen, *baked*	12 samples including chicken, chicken and ham, chicken and mushroom, and chicken and vegetable pies. 10.5-25% meat	15.8	6.26	5.58	2.06	2.19	0.23	1.12	0.03
314	**Cornish pastie**, *cooked*	10 samples, 5 brands	16.3	5.70	5.06	1.12	1.07	0.14	3.42	0.17
315	**Pork pie**, individual	8 samples of 8cm pies including buffet, Melton Mowbray. 28-39% meat	25.7	9.89	10.83	3.31	3.04	0.24	0.42	0.05
316	**Lamb samosa**, retail	8 samples. 20-23% meat	17.3	4.45	6.82	4.77	3.85	0.89	0.22	0.14
317	**Sausage roll**, flaky pastry, *cooked*	Mixed sample	29.0	10.92	11.25	2.35	2.40	0.25	2.96	0.17
318	**Spring rolls**, meat, takeaway	10 samples from different outlets	16.4	3.80	7.02	4.76	3.74	1.00	0.03	0.01
319	**Steak and kidney pie**, individual, *cooked*	Mixed sample	17.1	6.67	5.87	1.60	1.55	0.15	2.04	0.11
320	**Steak and kidney pudding**, canned	5 samples, different brands	11.6	5.01	3.95	0.63	0.60	0.21	0.83	0.39
Sausages										
321	**Beef sausages**, *raw*	6 samples of thick sausages. 50-55% meat	23.8	9.16	10.41	1.58	1.42	0.20	0.56	0.31
322	*fried in corn oil*	6 samples of thick sausages	19.7	7.25	8.78	1.73	1.51	0.24	0.31	0.20
323	*grilled* [a]	6 samples of thick sausages	19.5	7.69	8.35	1.37	1.26	0.15	0.38	0.22
324	**Pork sausages**, *raw*	10 samples, 7 brands of thick and thin sausages	22.7	8.23	9.73	3.07	2.77	0.26	0.11	0.02
325	*fried in corn oil*	16 samples	23.9	8.46	10.17	3.43	3.09	0.33	0.13	0.02
326	reduced fat, *raw*	7 samples, 5 brands of thick and thin sausages	10.6	3.72	4.44	1.59	1.46	0.13	0.06	0.02

[a] Contains 0.25g unidentified fatty acids per 100g food

Saturated fatty acids, g per 100g food

No.	Food	4:0	6:0	8:0	10:0	12:0	14:0	15:0	16:0	17:0	18:0	20:0	22:0	24:0
	Meat pies and pastries													
312	**Beef pie**, chilled/frozen, *baked*	0	0	0	0.01	0.03	0.40	0.04	5.71	0.08	2.02	0.11	0.08	0
313	**Chicken pie**, individual, chilled/frozen, *baked*	0	0	0	0.01	0.04	0.26	0.02	4.43	0.04	1.33	0.08	0.05	0
314	**Cornish pastie**, *cooked*	0	0	0	0	0.03	0.50	0.05	2.98	0.08	1.85	0.22	0	0
315	**Pork pie**, individual	0	0	0	0.02	0.02	0.51	0.05	6.00	0.10	3.11	0.07	0	0
316	**Lamb samosa**, retail	0	0	0	0.02	0.14	0.21	0.05	2.40	0.08	1.35	0.06	0.06	0.03
317	**Sausage roll**, flaky pastry, *cooked*	0	0	0	0	0.08	0.86	0.06	6.05	0.08	3.04	0.25	0.50	0
318	**Spring rolls**, meat, takeaway	0	0	0.12	0.11	0.90	0.40	0	1.58	0.02	0.54	0.07	0.05	0.03
319	**Steak and kidney pie**, individual, *cooked*	0	0	0	0	0.02	0.52	0.05	3.47	0.08	2.02	0.23	0.28	0
320	**Steak and kidney pudding**, canned	0	0	0	0	0	0.38	0.09	2.43	0.17	1.79	0.13	0.02	0
	Sausages													
321	**Beef sausages**, *raw*	0	0	0	0	0.02	0.53	0.07	5.49	0.16	2.87	0.02	0	0
322	*fried in corn oil*	0	0	0	0.02	0.02	0.44	0.06	4.34	0.13	2.23	0.02	0	0
323	*grilled*	0	0	0	0.02	0.02	0.47	0.05	4.57	0.13	2.39	0.04	0	0
324	**Pork sausages**, *raw*	0	0	0	0.02	0.02	0.34	0.02	5.10	0.08	2.60	0.04	0	0
325	*fried in corn oil*	0	0	0	0.02	0.02	0.33	0.02	5.21	0.09	2.72	0.04	0	0
326	*reduced fat, raw*	0	0	0	0.01	0.01	0.15	0.01	2.30	0.04	1.18	0.02	0	0

Meat products and dishes

Monounsaturated fatty acids, g per 100g food

No.	Food	10:1	12:1	14:1	cis 15:1	16:1	17:1	18:1	cis/trans 18:1 n-9	cis/trans 18:1 n-7	20:1	cis 22:1	cis/trans 22:1 n-11	cis/trans 22:1 n-9	cis 24:1	trans Monounsatd
Meat pies and pastries																
312	**Beef pie**, chilled/frozen, *baked*	0	0	0.03	0	0.27	0.06	6.15	6.75	0.57	0.20	0.06	0.05	0.05	0	1.20
313	**Chicken pie**, individual, chilled/frozen, *baked*	0	0	0.01	0	0.23	0.02	5.14	5.50	0.46	0.09	0.08	0.14	0	0	1.05
314	**Cornish pastie**, *cooked*	0	0	0.0E	0	0.37	0.08	4.27	5.44	1.15	0.30	Tr	Tr	Tr	0	3.32
315	**Pork pie**, individual	0	0	0	0	0.66	0.07	9.85	8.99	1.08	0.24	0.06	0	0	0	0.42
316	**Lamb samosa**, retail	0.01	0	0.02	0	0.15	0.05	6.35	N	N	0.18	0.06	N	N	0	0.22
317	**Sausage roll**, flaky pastry, *cooked*	0	0	0	0	0.69	0.06	9.59	9.01	1.44	0.30	0.61	0.58	0.72	0	2.65
318	**Spring rolls**, meat, takeaway	0.01	0	0	0	0.08	0.02	6.71	N	N	0.16	0.04	N	N	0	0.03
319	**Steak and kidney pie**, individual, *cooked*	0	0	0.02	0	0.33	0.07	4.68	4.24	1.35	0.44	0.34	0.51	0.16	0	1.94
320	**Steak and kidney pudding**, canned	0	0	0.10	0	0.35	0.10	3.12	3.00	0.39	0.21	0.07	0.06	0.06	0	0.65
Sausages																
321	**Beef sausages**, *raw*	0	0	0.16	0	0.87	0.16	9.12	8.63	0.85	0.11	0	0	0	0	0.51
322	*fried in corn oil*	0	0	0.13	0	0.72	0.11	7.70	7.20	0.63	0.13	0	0	0	0	0.29
323	*grilled*	0	0	0.15	0	0.75	N	7.36	7.05	0.58	0.09	0	0	0	0	0.35
324	**Pork sausages**, *raw*	0	0	0	0	0.61	0.08	8.82	8.20	0.70	0.21	0	0	0	0	0.11
325	*fried in corn oil*	0	0	0	0	0.62	0.09	9.24	8.66	0.69	0.22	0	0	0	0	0.13
326	*reduced fat, raw*	0	0	0	Tr	0.27	0.04	4.05	3.77	0.32	0.09	0	0	0	0	0.06

Meat products and dishes

Polyunsaturated fatty acids, g per 100g food

No.	Food	18:2	18:3	cis n-6 20:3	20:4	22:4	18:3	18:4	cis n-3 20:5	22:5	22:6	trans Polyunsatd
Meat pies and pastries												
312	**Beef pie**, chilled/frozen, *baked* a	1.63	0	0.01	0.02	0	0.25	0	0	0	0	0.02
313	**Chicken pie**, individual, chilled/frozen, *baked* a	1.79	0	0.01	0.02	0	0.21	0	0	0.01	0.01	0.07
314	**Cornish pastie**, *cooked*	1.06	0.02	0	0	0	0.14	0	0	0	0	0
315	**Pork pie**, individual b	2.89	0	0.02	0.05	0	0.22	0	0	0	0	0
316	**Lamb samosa**, retail a	3.83	0	0.01	0.01	Tr	0.81	0	0.01	0.02	0.04	0
317	**Sausage roll**, flaky pastry, *cooked* c	1.99	0	0	0	0	0.25	0	0	0	0	0.30
318	**Spring rolls**, meat, takeaway a	3.74	0	0	0.01	0	0.98	0	0	0.03	0	0
319	**Steak and kidney pie**, individual, *cooked* d	1.25	0	0	0.03	0	0.15	0	0	0	0	0.10
320	**Steak and kidney pudding**, canned a	0.31	0	0	0.04	0	0.18	0.03	0	0	0	0.18
Sausages												
321	**Beef sausages**, *raw* e	1.31	0	0	0.02	0	0.20	0	0	0	0	0.04
322	*fried in corn oil* e	1.45	0	0.02	0.02	0	0.24	0	0	0	0	0.02
323	*grilled* e	1.18	0	0	0	0	0.15	0	0	0	0	0.04
324	**Pork sausages**, *raw* f	2.62	0	0.02	0.04	0	0.25	0	0	0	0	0
325	*fried in corn oil* c	2.92	0	0.04	0.04	0	0.31	0	0	0	0	0
326	*reduced fat*, *raw* g	1.37	0	0.03	0.01	0	0.13	0	0	0	0	0

a Contains 0.02g 20:2 per 100g food
b Contains 0.07g 20:2 and 0.02g 20:4 n-3 per 100g food
c Contains 0.16g 20:2 per 100g food
d Contains 0.11g 20:2 per 100g food

e Contains 0.04g 20:2 per 100g food
f Contains 0.08g 20:2 per 100g food
g Contains 0.05g 20:2 per 100g food

Meat products and dishes

Fat and total fatty acids, g per 100g food

No.	Food	Description	Total fat	Satd	cis-Mono unsatd	Polyunsatd Total cis	n-6	n-3	Total trans	Total branched
Sausages *continued*										
327	**Pork sausages**, reduced fat, *fried in corn oil*	7 samples, 5 brands of thick and thin sausages	13.0	4.19	5.56	2.24	2.02	0.22	0.05	0.02
328	**Pork and beef sausages**, economy, *raw*	6 samples, 2 brands. 50-51% meat	20.5	7.39	8.87	2.55	2.41	0.21	0.11	0.10
329	**Premium sausages**, *raw*	10 samples, 9 brands including Cumberland and Lincolnshire sausages. 65-90% meat	18.7	6.81	7.73	2.72	2.46	0.26	0.09	0.03
330	*fried in vegetable oil*	10 samples, 9 brands	20.7	7.58	8.37	3.05	2.82	0.23	0.15	0.12
331	**Turkey sausages**, *raw*	1 packet	8.6	3.41	3.26	1.28	1.24	0.04	0.17	0.01
Continental-style sausages										
332	**Frankfurter**	10 samples, 7 brands of continental-style frankfurter. 75-90% meat	25.4	9.16	11.44	2.90	2.66	0.24	0.12	0.05
333	**Pepperami**	10 samples, 2 brands	51.1	19.39	22.83	5.01	4.63	0.38	0.29	0.14
334	**Salami**	11 samples, 6 brands including Danish, German, Italian and French salami	39.2	14.58	17.40	4.47	4.10	0.37	0.15	0
335	**Saveloy**, unbattered, takeaway	10 samples from fish and chip shops	27.9	9.39	11.68	4.46	4.04	0.47	0.55	0
Other meat products										
336	**Black pudding**, raw	8 samples, 6 brands	20.6	6.14	7.51	1.91	1.75	0.16	0.06	0.02
337	**Chicken in crumbs**, stuffed with cheese and vegetables, chilled/frozen, *baked*	7 samples, 5 brands. Fillings include cheese broccoli, chilli, and mushroom. 40-53% meat	13.9	3.99	4.29	4.21	3.87	0.45	0.28	0.08

Meat products and dishes

Saturated fatty acids, g per 100g food

No.	Food	4:0	6:0	8:0	10:0	12:0	14:0	15:0	16:0	17:0	18:0	20:0	22:0	24:0
Sausages continued														
327	**Pork sausages**, reduced fat, *fried in corn oil*	0	0	0	0	0.01	0.16	0.01	2.60	0.05	1.33	0.02	0	0
328	**Pork and beef sausages**, economy, *raw*	0	0	0	0	0.02	0.31	0.02	4.58	0.06	2.38	0.04	0	0
329	**Premium sausages**, *raw*	0	0	0	0.02	0.02	0.26	0.02	4.18	0.07	2.21	0.03	0	0
330	*fried in vegetable oil*	0	0	0	0.02	0.02	0.29	0.02	4.63	0.06	2.51	0.04	0	0
331	**Turkey sausages**, *raw*	0	0	0	0	0.01	0.11	0.02	2.42	0.02	0.84	0	0	0
Continental-style sausages														
332	**Frankfurter**	0	0	0	0.02	0.02	0.36	0.02	5.63	0.09	2.99	0.02	0	0
333	**Pepperami**	0	0	0	0	0.05	0.81	0.05	11.75	0.24	6.40	0.10	0	0
334	**Salami**	0	0	N	N	N	0.55	0	8.72	0.15	4.98	0	0	0.18
335	**Saveloy**, unbattered, takeaway	0	0	0	0.03	0.05	0.37	0	5.76	0.10	2.97	0.05	0.05	0
Other meat products														
336	**Black pudding**, *raw*	0	0	0	0.02	0.02	0.23	0.02	3.80	0.06	1.96	0.03	0	0
337	**Chicken in crumbs**, stuffed with cheese and vegetables, chilled/frozen, *baked*	0.04	0.05	0.04	0.08	0.10	0.34	0.04	2.52	0.04	0.74	0.01	0	0

Meat products and dishes

Monounsaturated fatty acids, g per 100g food

No.	Food	cis							cis/trans		cis		cis/trans		cis	trans
		10:1	12:1	14:1	15:1	16:1	17:1	18:1	18:1 n-9	18:1 n-7	20:1	22:1	22:1 n-11	22:1 n-9	24:1	Monounsatd
Sausages continued																
327	**Pork Sausages**, reduced fat, *fried in corn oil*	0	0	0	0	0.30	0.04	5.11	4.76	0.38	0.11	0	0	0	0	0.05
328	**Pork and beef sausages**, economy, *raw*	0	0	0	0	0.56	0	8.16	7.64	0.63	0.15	0	0	0	0	0.11
329	**Premium sausages**, *raw*	0	0	0	0	0.49	0.05	7.07	6.58	0.68	0.12	0	0	0	0	0.09
330	*fried in vegetable oil*	0	0	0	0	0.46	0	7.77	7.23	0.60	0.13	0	0	0	0	0.15
331	**Turkey sausages**, *raw*	0	0	0.02	0.02	0.43	0.01	2.77	2.74	0.15	0.02	0	0	0	0	0.17
Continental-style sausages																
332	**Frankfurter**	0	0	0.02	0	0.69	0.07	10.42	9.71	0.78	0.24	0	0	0	0	0.12
333	**Pepperami**	0	0	0.10	0	1.34	0.14	20.87	19.53	1.48	0.38	0	0	0	0	0.29
334	**Salami**	0	0	0	0	0.81	0	16.08	14.76	1.32	0.37	0.15	N	N	0	0.15
335	**Saveloy**, unbattered, *takeaway*	0	0	0	0	0.60	0.13	10.59	9.67	0.91	0.23	0.13	N	N	0	0.47
Other meat products																
336	**Black pudding**, *raw*	0	0	0	0	0.50	0.06	6.79	6.33	0.52	0.16	0	0	0	0	0.06
337	**Chicken in crumbs**, stuffed with cheese and vegetables, chilled/frozen, *baked*	0.01	0	0.03	0	0.27	0	3.95	3.78	0.31	0.03	0	0	0	0	0.17

Meat products and dishes

Polyunsaturated fatty acids, g per 100g food

No.	Food	cis n-6					cis n-3					trans
		18:2	18:3	20:3	20:4	22:4	18:3	18:4	20:5	22:5	22:6	Polyunsatd
Sausages *continued*												
327	**Pork sausages**, reduced fat, *fried in corn oil* a	1.94	0	0	0.04	0	0.22	0	0	0	0	0
328	**Pork and beef sausages**, economy, *raw*	2.30	0	0	0.04	0	0.21	0	0	0	0	0
329	**Premium sausages**, *raw* b	2.30	0	0.05	0.05	0	0.23	0	0	0	0	0
330	*fried in vegetable oil* c	2.64	0	0.02	0.06	0	0.23	0	0	0	0	0
331	**Turkey sausages**, *raw*	1.24	0	0	0	0	0.04	0	0	0	0	0
Continental-style sausages												
332	**Frankfurter** b	2.49	0	0.02	0.07	0	0.21	0	0	0	0	0
333	**Pepperami** d	4.35	0	0.10	0	0	0.38	0	0	0	0	0
334	**Salami**	4.10	0	0	0	0	0.37	0	0	0	0	0
335	**Saveloy**, unbattered, takeaway	3.99	0	0	0	0	0.47	0	0	0	0	0.08
Other meat products												
336	**Black pudding**, *raw* e	1.68	0	0.02	0.02	0	0.14	0	0	0	0	0
337	**Chicken in crumbs**, stuffed with cheese and vegetables, chilled/frozen, *baked* f	3.77	0	0	0.01	0	0.39	0.01	0.01	0	0	0.11

a Contains 0.05g 20:2 per 100g food
b Contains 0.09g 20:2 per 100g food
c Contains 0.10g 20:2 per 100g food
d Contains 0.19g 20:2 per 100g food
e Contains 0.06g 20:2 per 100g food
f Contains 0.01g 16:4 per 100g food

Meat products and dishes

Fat and total fatty acids, g per 100g food

No.	Food	Description	Total fat	Satd	cis-Mono unsatd	Polyunsatd Total cis	n-6	n-3	Total trans	Total branched
	Other meat products continued									
338	**Chicken kiev**, frozen, *baked*	5 samples, 4 brands. 45-60% meat	16.9	6.95	5.11	3.31	2.93	0.41	0.40	0.10
339	**Chicken roll**, *cooked*	10 samples, 3 brands	4.8	1.50	2.04	0.90	0.79	0.13	0.08	0.02
340	**Chicken tandoori**, chilled, *reheated*	7 samples, 6 brands. 95-96% meat	10.8	3.21	4.93	1.93	1.77	0.17	0.07	0.07
341	**Chicken tikka**, chilled, *reheated*	7 samples, different brands	9.8	3.18	3.77	1.91	1.58	0.42	0.29	0.08
342	**Corned beef**	10 samples, 4 brands	10.9	5.45	3.77	0.18	0.28	0.03	0.68	0.22
343	**Doner kebabs**, *cooked*, lean only	6 samples from different outlets	31.4	14.32	9.49	0.67	0.90	0.45	2.43	0.65
344	**Faggots in gravy**	10 samples	7.5	2.73	3.06	1.08	0.93	0.17	0.14	0.03
345	**Ham and pork**, chopped, canned	10 samples	22.4	8.35	10.63	2.22	N	N	N	N
346	**Lamb roasts**, frozen, *cooked*	2 samples. 90% meat	13.3	6.23	4.10	0.50	0.34	0.20	0.97	0.24
347	**Luncheon meat**, canned	10 samples, 9 brands	23.8	8.47	10.66	2.62	2.53	0.38	0.40	N
348	**Meat loaf**, chilled/frozen, *reheated*	10 samples, 5 brands	15.8	5.54	7.03	1.89	1.73	0.16	0.13	0.01
349	**Mince in gravy**, canned	10 samples, 6 brands; beef	11.7	4.83	4.85	0.34	0.38	0.09	0.57	0.10
350	**Pâté**, liver	20 samples including canned	32.7	9.37	11.78	2.94	2.70	0.24	0.02	0.01
351	meat, low fat	11 samples, assorted types; pork meat and liver based	12.0	3.37	3.86	1.43	1.32	0.11	0.08	0.03
352	**Pork haslet**, *cooked*	10 samples, 6 brands. 69% meat	12.6	4.41	4.88	2.02	1.80	0.18	0.35	0.05
353	**Pork roasts**, frozen, *cooked*	2 samples. 88% meat	12.9	4.70	5.46	1.73	1.62	0.12	0.07	0.05
354	**Rissoles**, savoury, *cooked*	10 samples	16.7	7.05	6.57	0.52	0.50	0.11	0.92	0.30
355	**Shish kebabs**, *cooked*, lean only	6 samples from different outlets	10.0	3.68	3.77	0.72	0.52	0.27	0.58	0.15
356	**Tongue slices**	8 samples, 6 brands	14.0	5.57	5.74	0.74	0.65	0.21	0.68	0.29
357	**Turkey roasts**, frozen, *cooked*	2 samples. 76% meat	7.0	2.51	2.94	1.03	0.96	0.07	0.09	0.03

Meat products and dishes

Saturated fatty acids, g per 100g food

No.	Food	4:0	6:0	8:0	10:0	12:0	14:0	15:0	16:0	17:0	18:0	20:0	22:0	24:0
	Other meat products continued													
338	**Chicken kiev**, frozen, *baked*	0.11	0.14	0.09	0.18	0.24	0.84	0.10	3.61	0.07	1.49	0.04	0.02	0
339	**Chicken roll**, *cooked*	0	0	0	0	0.01	0.05	0.01	1.10	0.01	0.30	0.01	0.01	0
340	**Chicken tandoori**, chilled, *reheated*	0	0	0	0.01	0.01	0.11	0.02	2.38	0.02	0.65	0.01	0	0
341	**Chicken tikka**, chilled, *reheated*	0.08	0.06	0.07	0.10	0.28	0.40	0.04	1.51	0.03	0.58	0.03	0.01	0
342	**Corned beef**	0	0	0	0	0	0	0.38	0.06	3.17	0.14	1.65	0.05	0
343	**Doner kebabs**, *cooked*, lean only	0	0	0	0.06	0.13	1.37	0.22	6.23	0.40	5.88	0.04	0	0
344	**Faggots in gravy**	0	0	0	Tr	0.01	0.11	0.01	1.70	0.03	0.84	0.02	0	0
345	**Ham and pork**, chopped, canned	0	0	0	0	0	0.34	0	5.34	Tr	2.67	0	0	0
346	**Lamb roasts**, frozen, *cooked*	0	0	0	0.02	0.04	0.45	0.09	2.71	0.17	2.72	0.02	0	0
347	**Luncheon meat**, canned	0	0	0	0.02	0.02	0.31	0	4.79	0.18	2.88	0.27	0	0
348	**Meat loaf**, chilled/frozen, *reheated*	0	0	0	0.03	0.03	0.24	0.01	3.23	0.06	1.83	0.10	0	0
349	**Mince in gravy**, canned	0	0	0	0	0	0.33	0.05	2.62	0.11	1.71	0.01	0	0
350	**Pâté**, liver	0	0	0	0	0.06	0.36	0.01	5.66	0.10	3.15	0.02	0	0
351	meat, low fat	0	0	0	0.01	0.01	0.04	0.01	2.06	0.04	1.20	0.01	0	0
352	**Pork haslet**, *cooked*	0	0	0	0.01	0.01	0.18	0.01	2.73	0.01	1.42	0.04	0	0
353	**Pork roasts**, frozen, *cooked*	0	0	0	0.01	0.01	0.18	0.01	2.90	0.05	1.53	0.01	0	0
354	**Rissoles**, savoury, *cooked*	0	0	0	0.03	0.02	0.48	0.08	3.67	0.17	2.59	0.02	0	0
355	**Shish kebabs**, *cooked*, lean only	0	0	0	0.01	0.02	0.29	0.05	1.81	0.09	1.39	0.01	0	0
356	**Tongue slices**	0	0	0	0.01	0.01	0.36	0.09	2.87	0.18	2.03	0.02	0	0
357	**Turkey roasts**, frozen, *cooked*	0	0	0	0	0.01	0.09	0.02	1.81	0.03	0.56	0.01	0	0

Meat products and dishes

Monounsaturated fatty acids, g per 100g food

No.	Food	cis 10:1	12:1	14:1	15:1	16:1	17:1	18:1	cis/trans 18:1 n-9	18:1 n-7	cis 20:1	22:1	cis/trans 22:1 n-11	22:1 n-9	cis 24:1	trans Monounsatd
	Other meat products continued															
338	**Chicken kiev**, frozen, *baked*	0	0	0.03	0	0.23	0.05	4.70	4.43	0.26	0.08	0.02	0.02	0	0	0.37
339	**Chicken roll**, *cooked*	0	0	0.01	0	0.24	0	1.73	1.66	0.12	0.05	0	0	0	0	0.07
340	**Chicken tandoori**, chilled, *reheated*	0	0	0.02	0	0.57	0.02	4.26	4.03	0.28	0.06	0	0	0	0	0.07
341	**Chicken tikka**, chilled, *reheated*	0.01	0	0.03	0	0.15	0.02	3.49	3.35	0.27	0.06	0.01	0	0.01	0	0.20
342	**Corned beef**	0	0	0.07	0	0.38	0.08	3.21	3.18	0.48	0.02	0	0	0	0	0.54
343	**Doner kebabs**, *cooked*, lean only	0	0	0.04	0.01	0.53	0.19	8.66	8.38	1.94	0.06	0	0	0	0	2.24
344	**Faggots in gravy**	0	0	0	0	0.19	0	2.81	2.66	0.24	0.06	0	0	0	0	0.13
345	**Ham and pork**, chopped, canned	0	0	0	0	0.73	Tr	9.73	N	N	0.17	0	0	0	0	N
346	**Lamb roasts**, frozen, *cooked*	0	0	0.02	0	0.17	0	3.88	3.65	0.92	0.02	0	0	0	0	0.93
347	**Luncheon meat**, canned	0	0	0	0	0.71	0.18	9.40	8.38	1.00	0.38	0	0	0	0	0.11
348	**Meat loaf**, chilled/frozen, *reheated*	0.03	0	0	0	0.40	0.06	6.22	5.80	0.49	0.32	0	0	0	0	0.13
349	**Mince in gravy**, canned	0	0	0.10	0	0.43	0.10	4.20	3.70	0.86	0.03	0	0	0	0	0.44
350	**Pâté**, liver	0	0	0	0	0.68	0.10	10.77	9.95	0.76	0.23	0	0	0	0	0
351	meat, low fat	0	0	0	0	0.22	0.03	3.53	3.29	0.28	0.08	0	0	0	0	0.08
352	**Pork haslet**, *cooked*	0	0	0	0	0.07	0.02	4.67	4.38	0.38	0.12	0	0	0	0	0.35
353	**Pork roasts**, frozen, *cooked*	0	0	0	0	0.32	0.02	5.00	4.64	0.41	0.11	0	0	0	0	0.07
354	**Rissoles**, savoury, *cooked*	0	0	0.14	0	0.58	0.11	5.68	5.62	0.64	0.06	0	0	0	0	0.83
355	**Shish kebabs**, *cooked*, lean only	0	0	0.01	0	0.15	0.07	3.51	3.86	0.06	0.03	0	0	0	0	0.50
356	**Tongue slices**	0	0	0.09	Tr	0.42	0.16	5.04	4.91	0.49	0.03	0	0	0	0	*0.57*
357	**Turkey roasts**, frozen, *cooked*	0	0	0.02	0.01	0.46	0.01	2.42	2.28	0.19	0.03	0	0	0	0	0.09

Meat products and dishes

Polyunsaturated fatty acids, g per 100g food

No.	Food	cis n-6					18:3	18:4	cis n-3			trans Polyunsatd
		18:2	18:3	20:3	20:4	22:4			20:5	22:5	22:6	Polyunsatd
	Other meat products continued											
338	**Chicken kiev**, frozen, baked	2.91	0	Tr	Tr	0	0.37	0	0.02	0	0	0.03
339	**Chicken roll**, cooked a	0.71	0	0	0.07	0	0.05	0	0	0.03	0.05	0.01
340	**Chicken tandoori**, chilled, reheated a	1.70	0	0.02	0.03	0	0.16	0	0	0	0	0
341	**Chicken tikka**, chilled, reheated a	1.53	0	Tr	0.01	0	0.33	0	0.01	0.01	0.01	0.08
342	**Corned beef**	0.15	0	0	0	0	0.03	0	0	0	0	0.14
343	**Doner kebabs**, cooked, lean only	0.30	0.17	0	0.03	0	0.31	0	0.02	0.06	0	0.19
344	**Faggots in gravy** b	0.85	0	0.02	0.05	0	0.11	0	0.01	0.02	0.01	0.01
345	**Ham and pork**, chopped, canned	2.05	0	0	0.06	0	0.11	0	0	0	0	N
346	**Lamb roasts**, frozen, cooked	0.29	0	0.01	0	0	0.20	0	0	0	0	0.04
347	**Luncheon meat**, canned c	2.06	0	0	0.07	0	0.38	0	0	0	0	0.29
348	**Meat loaf**, chilled/frozen, reheated d	1.48	0.01	0.01	0.06	0	0.12	0	0	0.04	0	0
349	**Mince in gravy**, canned	0.23	0.02	0	0	0	0.09	0	0	0	0	0.13
350	**Pâté**, liver e	2.41	0	0	0.17	0	0.18	0	0.04	0.04	0.02	0.02
351	**meat**, low fat f	1.16	0	0.02	0.12	0	0.10	0	0	0	0	0
352	**Pork haslet**, cooked g	1.73	0	0.04	0.05	0	0.15	0	0	0	0	0
353	**Pork roasts**, frozen, cooked g	1.49	0	0.01	0.06	0	0.11	0	0	0	0	0
354	**Rissoles**, savoury, cooked	0.39	0.02	0	0	0	0.11	0	0	0	0	0.09
355	**Shish kebabs**, cooked, lean only h	0.37	0	0	0.03	0	0.18	0.01	0.02	0.03	0.01	0.08
356	**Tongue slices** a	0.47	0	0.02	0.07	0	0.11	0	0.02	0.04	Tr	0.12
357	**Turkey roasts**, frozen, cooked	0.93	0	0	0.03	0	0.07	0	0	0	0	0.01

a Contains 0.01g 20:2 per 100g food
b Contains 0.03g 20:2 per 100g food
c Contains 0.11g 20:2 per 100g food
d Contains 0.07g 20:2 per 100g food
e Contains 0.12g 20:2 per 100g food
f Contains 0.04g 20:2 per 100g food
gContains 0.06g 20:2 per 100g food
h Contains 0.01g 22:2 per 100g food

Meat products and dishes

Fat and total fatty acids, g per 100g food

No.	Food	Description	Total fat	Satd	cis-Mono unsatd	Polyunsatd Total cis	n-6	n-3	Total trans	Total branched
	Meat dishes									
358	**Beef stew**, meat only	6 samples, 5 brands	3.8	1.77	1.32	0.14	0.17	0.05	0.18	0.07
359	**Beef in sauce with vegetables**, chilled/frozen, *reheated*	8 samples, 5 brands including braised steak, beef bourguignonne and goulash	4.6	1.48	1.78	0.78	0.77	0.09	0.17	0.05
360	**Beef stir-fried with green peppers in black bean sauce**, takeaway	10 samples from different outlets	5.6	1.26	3.35	0.50	0.35	0.15	0.01	0.08
361	**Cannelloni**, chilled/frozen, *reheated*	10 samples, 4 brands	5.0	2.01	1.82	0.56	0.53	0.06	0.27	0.06
362	**Chicken chop suey**, takeaway	10 samples from different outlets	4.7	0.83	2.30	1.26	0.95	0.28	0.02	0.02
363	**Chicken fried rice**, takeaway	10 samples from different outlets	6.0	0.99	3.01	1.57	1.18	0.38	0.02	0.01
364	**Chicken in sauce with vegetables**, chilled/frozen, *reheated*	10 samples including chicken, tomato and mushroom casserole and chicken creole	5.1	1.27	2.25	1.10	0.99	0.17	0.07	0.02
365	**Chicken in white sauce**, canned	10 samples, 4 brands	8.3	2.32	3.85	1.61	1.54	0.10	0.03	0.02
366	**Chicken korma**, takeaway	10 samples from different outlets	14.6	5.66	5.22	2.77	2.18	0.56	0.10	0.11
367	**Chicken tikka masala**, takeaway	10 samples from different outlets	11.5	3.85	4.40	2.44	1.94	0.43	0.06	Tr
368	**Chicken with cashew nuts**, takeaway	10 samples from different outlets	8.7	1.45	4.65	2.14	1.79	0.34	0.01	0.01
369	**Cottage/shepherd's pie**, chilled/frozen, *reheated*	11 samples including beef and lamb. 11-25% meat	5.4	2.35	2.01	0.38	0.36	0.06	0.28	0.10
370	**Lamb/beef hotpot with potatoes**, chilled/frozen, *reheated*	10 samples, 6 brands of beef and Lancashire hotpot. 10-32% meat	4.4	1.62	1.58	0.45	0.39	0.12	0.35	0.05
371	**Irish stew**, canned	10 samples, 2 brands	5.1	2.40	1.66	0.24	0.20	0.06	0.37	0.11
372	**Lamb rogan josh**, takeaway	10 samples from different outlets	10.0	2.65	4.22	2.27	1.79	0.45	0.07	0.09
373	**Lasagne**, chilled/frozen, *reheated*	12 samples, 11 brands. 10-20% meat	6.1	2.72	2.03	0.65	0.59	0.08	0.28	N

Saturated fatty acids, g per 100g food

No.	Food	4:0	6:0	8:0	10:0	12:0	14:0	15:0	16:0	17:0	18:0	20:0	22:0	24:0
Meat dishes														
358	**Beef stew**, meat only	0	0	0	0	0	0.10	0.02	0.89	0.05	0.69	0.01	0	0
359	**Beef in sauce with vegetables,** chilled/frozen, *reheated*	0	0	0	0.01	0.01	0.08	0.01	0.83	0.03	0.48	0.01	0	Tr
360	**Beef stir-fried with green peppers in black bean sauce,** takeaway	0	0.03	0.01	Tr	Tr	0.03	0.05	0.68	0.02	0.29	0.04	0.06	0.03
361	**Cannelloni,** chilled/frozen, *reheated*	0	0	0	0.04	0.06	0.17	0.02	1.06	0.03	0.61	0.01	0.01	0
362	**Chicken chop suey,** takeaway	0	0	0.02	0	0	0.01	0.02	0.51	0.01	0.17	0.02	0.03	0.02
363	**Chicken fried rice,** takeaway	0	0	0.02	0	0	0.02	0.02	0.66	0.01	0.20	0.03	0.02	0.02
364	**Chicken in sauce with vegetables,** chilled/frozen, *reheated*	0	0.01	0.01	0.01	0.04	0.07	0.01	0.82	0.01	0.27	0.01	0	Tr
365	**Chicken in white sauce,** canned	0	0	0	0	0.01	0.08	0.02	1.69	0.02	0.49	0.02	0	Tr
366	**Chicken korma,** takeaway	0.14	0.06	0.15	0.18	0.86	0.78	0.07	2.40	0.05	0.84	0.05	0.03	0.05
367	**Chicken tikka masala,** takeaway	0	0.03	0.12	0.11	0.53	0.48	0.06	1.74	0.03	0.59	0.04	0.04	0.08
368	**Chicken with cashew nuts,** takeaway	0	0	0.01	0	0	0	0	0.84	0.01	0.49	0.04	0.02	0.01
369	**Cottage/shepherd's pie,** chilled/frozen, *reheated*	0.02	0.01	0.01	0.02	0.04	0.21	0.03	1.19	0.05	0.74	0.01	0.01	0
370	**Lamb/beef hotpot with potatoes,** chilled/frozen, *reheated*	0	0	0	0	0.01	0.07	0.02	0.92	0.04	0.55	0.01	0	0
371	**Irish stew,** canned	0	0	0	0	0.01	0.14	0.03	1.12	0.07	1.02	0.01	0	0
372	**Lamb rogan josh,** takeaway	0	0	0.04	0.02	0.06	0.14	0.05	1.46	0.04	0.67	0.04	0.06	0.08
373	**Lasagne,** chilled/frozen, *reheated*	0.06	0.04	0.03	0.06	0.10	0.31	0.04	1.35	0.04	0.66	0.01	0	0

Meat products and dishes

Monounsaturated fatty acids, g per 100g food

No. Food	cis							cis/trans		cis		cis/trans		cis	trans
	10:1	12:1	14:1	15:1	16:1	17:1	18:1	18:1 n-9	18:1 n-7	20:1	22:1	22:1 n-11	22:1 n-9	24:1	Monounsatd
Meat dishes															
358 **Beef stew**, meat only	0	0	0	0	0.10	0	1.21	1.16	0.14	0.01	0	0	0	0	0.13
359 **Beef in sauce with vegetables**, chilled/frozen, *reheated*	0	0	0	0	0.09	0	1.68	1.65	0.13	0.02	0	0	0	0	0.13
360 **Beef stir-fried with green peppers in black bean sauce**, takeaway	0.01	0	0.01	0	0.05	0.01	2.68	N	N	0.46	0.13	N	N	0	0.01
361 **Cannelloni**, chilled/frozen, *reheated*	0	0	0.03	0	0.12	0.02	1.62	1.59	0.18	0.03	0	0	N	0	0.24
362 **Chicken chop suey**, takeaway	0	0	0	0	0.04	0.01	2.02	N	N	0.18	0.05	N	N	0	0.01
363 **Chicken fried rice**, takeaway	0	0	0	0	0.08	0.01	2.76	N	N	0.13	0.02	N	N	0	0.01
364 **Chicken in sauce with vegetables**, chilled/frozen, *reheated*	0	0	0	0	0.14	0	2.08	1.99	0.13	0.02	0	0	0	0	0.06
365 **Chicken in white sauce**, canned	0	0	0.01	0	0.39	0.02	3.39	3.19	0.18	0.05	0	0	0	0	0
366 **Chicken korma**, takeaway	0.02	0.01	0.04	0	0.13	0.03	4.87	N	N	0.09	0.03	N	N	0	0.10
367 **Chicken tikka masala**, takeaway	0.01	0	0.02	0	0.09	0.02	4.12	N	N	0.08	0.05	N	N	0	0.06
368 **Chicken with cashew nuts**, takeaway	0	0	0	0	0.06	0.01	4.51	N	N	0.06	0.01	N	N	0	0
369 **Cottage/shepherd's pie**, chilled/frozen, *reheated*	0	0	0.04	0	0.15	0.03	1.75	1.90	0.04	0.02	0.01	0.01	0.01	0	0.23
370 **Lamb/beef hotpot with potatoes**, chilled/frozen, *reheated*	0	0	0.01	0	0.05	0.02	1.49	1.52	0.21	0.01	0	0	0	0	0.29
371 **Irish stew**, canned	0	0	Tr	0	0.08	0.04	1.52	0.83	0.17	0.02	0	0	0	0	0.35
372 **Lamb rogan josh**, takeaway	0	0	0.01	0	0.08	0.03	3.73	N	N	0.29	0.08	N	N	0	0.07
373 **Lasagne**, chilled/frozen, *reheated*	0	0	0.04	0	0.12	0.02	1.82	1.78	0.19	0.02	0	0	0	0	0.23

Meat products and dishes

Polyunsaturated fatty acids, g per 100g food

| No. | Food | cis n-6 | | | | | 18:3 | 18:4 | cis n-3 | | | trans Polyunsatd |
		18:2	18:3	20:3	20:4	22:4			20:5	22:5	22:6	Polyunsatd
	Meat dishes											
358	**Beef stew**, meat only	0.12	0	Tr	Tr	0	0.03	0	0	0	0	0.04
359	**Beef in sauce with vegetables**, chilled/frozen, *reheated*	0.71	0.01	0.01	0.02	0	0.06	0	0.01	0.01	Tr	0.04
360	**Beef stir-fried with green peppers in black bean sauce**, takeaway	0.35	0	0	0.01	0	0.05	0	0.04	0.02	0.04	0
361	**Cannelloni**, chilled/frozen, *reheated*	0.51	0	0	0	0	0.06	0	0	0	0	0.03
362	**Chicken chop suey**, takeaway [a]	0.92	0	0.02	0.02	0	0.22	0	0.01	0.01	0.03	0.01
363	**Chicken fried rice**, takeaway [b]	1.11	0	0.01	0.02	0.05	0.32	0	0.01	0.04	0.02	0.01
364	**Chicken in sauce with vegetables**, chilled/frozen, *reheated*	0.96	0.00	Tr	0.02	0	0.15	0	Tr	0.01	0.01	0.01
365	**Chicken in white sauce**, canned	1.35	0.02	0.05	0.09	0	0.07	0	0	0.01	0.02	0.03
366	**Chicken korma**, takeaway [a]	2.14	0.03	0.01	Tr	0.01	0.50	0	0.01	0.03	0	0
367	**Chicken tikka masala**, takeaway [c]	1.92	0	0.03	0	0	0.38	0	0.01	0.03	0	0
368	**Chicken with cashew nuts**, takeaway [b]	1.77	0	Tr	0.02	0.00	0.32	0	0	0.02	0	0.01
369	**Cottage/shepherd's pie**, chilled/frozen, *reheated*	0.31	0	0	0.01	0	0.06	0	0	0.01	0	0.04
370	**Lamb/beef hotpot with potatoes**, chilled/frozen, *reheated*	0.33	0	0	0.01	0	0.10	0	0	0.01	0	0.06
371	**Irish stew**, canned	0.17	0	0	0.01	0	0.06	0	0	0	0	0.02
372	**Lamb rogan josh**, takeaway [a]	1.75	0.02	0.01	0.01	0	0.39	0	0.02	0.02	0	0
373	**Lasagne**, chilled/frozen, *reheated*	0.55	0	0	0.01	0	0.08	0	0	0.01	0	0.05

[a] Contains 0.03g 20:2 per 100g food
[b] Contains 0.01g 20:2 per 100g food
[c] Contains 0.08g 20:2 per 100g food

Meat products and dishes

Fat and total fatty acids, g per 100g food

No.	Food	Description	Total fat	Satd	cis-Mono unsatd	Polyunsatd Total cis	n-6	n-3	Total trans	Total branched
	Meat dishes *continued*									
374	**Moussaka**, chilled/frozen/longlife, *reheated*	8 samples, 4 brands of beef and lamb. 20-23% meat	8.3	2.85	3.36	1.01	0.86	0.29	0.40	0.10
375	**Pancakes, beef**, frozen, *shallow-fried in vegetable oil*	Coated crispy pancakes. 4 samples, 2 brands. 15-16% meat	15.7	2.37	7.02	5.13	4.41	0.83	0.13	0.01
376	**chicken**, frozen, *shallow-fried in vegetable oil*	Coated crispy pancakes. 4 samples, 2 brands. 15-16% meat	14.2	1.58	6.53	5.10	4.32	0.84	0.08	0.03
377	**Pork spare-ribs**, barbecue style, chilled/frozen, *reheated*	5 samples including American and Chinese style, and hot and spicy ribs. 80-90% meat	17.1	6.17	6.73	2.71	2.41	0.31	0.12	0.08
378	**Spaghetti bolognese**, chilled, *reheated*	12 samples, sauce only. 10-18% meat	5.7	2.27	2.35	0.42	0.33	0.13	0.24	0.11
379	**Sweet and sour pork**, battered, takeaway	10 samples from different outlets, sauce not included	13.9	2.25	7.01	3.57	2.70	0.84	0.04	N
380	**Tagliatelle with ham**, mushrooms and cheese, chilled/frozen/longlife, *reheated*	11 samples, 3 brands	7.1	3.71	1.96	0.49	0.46	0.08	0.38	0.10

Meat products and dishes

Saturated fatty acids, g per 100g food

No.	Food	4:0	6:0	8:0	10:0	12:0	14:0	15:0	16:0	17:0	18:0	20:0	22:0	24:0
Meat dishes continued														
374	**Moussaka**, chilled/frozen/longlife, *reheated*	0.03	0.04	0.02	0.06	0.07	0.33	0.04	1.36	0.05	0.80	0.03	0	0.01
375	**Pancakes, beef**, frozen, *shallow-fried in vegetable oil*	0	0	0	0	0.01	0.07	0.01	1.60	0.03	0.59	0.04	0	0
376	**chicken**, frozen, *shallow-fried in vegetable oil*	0	0	0	0	0	0.03	0	1.12	0.01	0.36	0.05	0	0
377	**Pork spare-ribs**, barbecue style, chilled/frozen, *reheated*	0	0	0	0.01	0.02	0.22	0.01	3.71	0.06	2.11	0.03	0	0
378	**Spaghetti bolognese**, chilled, *reheated*	0	0	Tr	0	Tr	0.13	0.03	1.19	0.06	0.83	0.01	Tr	0
379	**Sweet and sour pork**, battered, takeaway	0	0	0.01	0	0.01	0.05	0.01	1.42	0.02	0.58	0.07	0.06	0.03
380	**Tagliatelle with ham**, mushrooms and cheese, chilled/frozen/longlife, *reheated*	0.14	0.09	0.06	0.13	0.16	0.51	0.05	1.84	0.04	0.67	0.01	0	0.01

Meat products and dishes

No.	Food	cis 10:1	12:1	14:1	15:1	16:1	17:1	18:1	cis/trans 18:1 n-9	18:1 n-7	cis 20:1	22:1	cis/trans 22:1 n-11	22:1 n-9	cis 24:1	trans Monounsatd
Meat dishes continued																
374	**Moussaka**, chilled/frozen/longlife, *reheated*	0.01	0	0.02	0	0.08	0.03	3.16	3.08	0.29	0.04	0.01	0	0.01	0.01	0.27
375	**Pancakes, beef**, frozen, *shallow-fried in vegetable oil*	0	0	0.03	0	0.18	0.03	6.72	6.37	0.37	0.06	0	0	0	0	0.07
376	**chicken**, frozen, *shallow-fried in vegetable oil*	0	0	0	0	0.08	N	6.33	5.99	0.34	0.12	0	0	0	0	0.03
377	**Pork spare-ribs**, barbecue style, chilled/frozen, *reheated*	0	0	0	0	0.39	0	6.16	5.70	0.50	0.15	0.01	Tr	Tr	0.01	0.09
378	**Spaghetti bolognese**, chilled, *reheated*	0	0	0.03	0	0.16	0.04	2.10	2.01	0.22	0.02	0	0	0	0	0.20
379	**Sweet and sour pork**, battered, takeaway	0.01	0	0	0	0.10	0.02	6.67	N	N	0.18	0.04	N	N	0	0.02
380	**Tagliatelle with ham**, mushrooms and cheese, chilled/frozen/longlife, *reheated*	0.01	Tr	0.04	Tr	0.10	0.02	1.75	1.79	0.21	0.02	0.01	Tr	Tr	0	0.32

Meat products and dishes

Polyunsaturated fatty acids, g per 100g food

No.	Food	cis n-6					cis n-3					trans
		18:2	18:3	20:3	20:4	22:4	18:3	18:4	20:5	22:5	22:6	Polyunsatd
Meat dishes *continued*												
374	**Moussaka**, chilled/frozen/longlife, *reheated*	*0.76*	*0*	*0*	*0.01*	*0*	*0.22*	*0*	*0.01*	*0.01*	*0*	*0.13*
375	**Pancakes**, beef, frozen, *shallow-fried in vegetable oil*	4.36	0	0.04	0.03	0	0.70	0	0	0	0	0.06
376	**chicken**, frozen, *shallow-fried in vegetable oil*	4.27	0	0	0.04	0	0.78	0	0	0	0	0.05
377	**Pork spare-ribs**, barbecue style, chilled/frozen, *reheated* [a]	2.21	0.01	0.05	0.09	0	0.19	0	0.02	0.05	0	0.03
378	**Spaghetti bolognese**, chilled, *reheated*	*0.28*	*0.02*	*0*	*0.02*	*0*	*0.11*	*0*	*0.01*	*0.01*	*0*	*0.04*
379	**Sweet and sour pork**, battered, takeaway [b]	2.66	0	0	0.02	0.02	0.79	0	0	0.03	0.02	0.02
380	**Tagliatelle with ham**, mushrooms and cheese, chilled/frozen/longlife, *reheated*	*0.39*	*0*	*0*	*0.01*	*0*	*0.07*	*0*	*Tr*	*0.01*	*0*	*0.06*

[a] Contains 0.01g 16:4, 0.08 20:2 per 100g food
[b] Contains 0.02g 20:2 per 100g food

Fish and fish products

Fat and total fatty acids, g per 100g food

No.	Food	Description	Total fat	Satd	cis-Mono unsatd	Polyunsatd Total cis	n-6	n-3	Total trans	Total branched
White Fish										
381	**Cod**, *raw*	Samples from 3 different shops	0.7	0.13	0.08	0.28	0.02	0.26	0	0
382	**Haddock**, *raw*	Mixed sample; fillets only	0.6	0.12	0.09	0.20	0.03	0.17	Tr	0
383	**Plaice**, *raw*	8 samples and literature sources	1.4	0.23	0.39	0.35	0.01	0.32	Tr	0.02
Fatty fish										
384	**Eel, jellied** [a]	10 samples, including jelly	7.1	1.89	3.50	0.96	0.32	0.64	Tr	0.05
385	**Herring**, *raw*	35 fish purchased whole over the year	13.2	3.65	5.98	2.17	0.32	1.83	Tr	0.11
386	**Kippers**, *raw*	10 samples from assorted outlets	17.7	2.76	9.26	3.86	0.24	3.52	0	0.06
387	**Mackerel**, *raw*	10 samples from assorted outlets, purchased whole; flesh and skin	16.1	3.23	7.87	3.27	0.51	2.78	0	0.06
388	**Pilchards**, canned in tomato sauce	10 samples; whole contents	8.1	1.71	2.22	3.36	0.28	2.97	0	0.01
389	**Salmon**, canned in brine, drained	14 samples, 8 brands; red and pink	7.8	1.38	3.29	2.24	0.27	1.85	0	0.05
390	**Sardines**, canned in oil, drained	13 samples, 10 brands	14.1	2.88	4.76	4.94	2.60	2.27	0.11	0
391	**Sardines**, canned in tomato sauce	10 samples, 8 brands; whole contents	9.9	2.79	2.90	3.22	1.04	2.11	0	0
392	**Sprats**, *raw*	10 samples from assorted outlets, purchased whole	11.0	2.15	4.67	3.04	0.31	2.68	0	0.04
393	**Trout, rainbow**, *raw*	11 samples from assorted outlets, purchased whole	5.2	1.10	1.85	1.73	0.41	1.32	0	
394	*grilled*, flesh only	11 samples from assorted outlets	5.4	1.12	2.02	1.72	0.46	1.25	0	0
395	**Tuna**, canned in brine, drained	10 samples, 9 brands	0.6	0.23	0.10	0.21	0.04	0.17	0	0
396	canned in oil, drained	10 samples, 6 brands; skipjack tuna	9.0	1.43	2.13	4.32	3.29	1.22	0.23	0

[a] Contains 0.04g unidentified fatty acids per 100g food

Fish and fish products

Saturated fatty acids, g per 100g food

No.	Food	4:0	6:0	8:0	10:0	12:0	14:0	15:0	16:0	17:0	18:0	20:0	22:0	24:0
White fish														
381	**Cod**, *raw*	0	0	0	0	0	0	0	0.11	0	0.02	0	0	0
382	**Haddock**, *raw*	0	0	0	0	0	0.01	0	0.08	0.01	0.03	0	0	0
383	**Plaice**, *raw*	0	0	0	0	0	0.03	Tr	0.16	Tr	0.03	0	0	0
Fatty fish														
384	**Eel, jellied**	0	0	0	0	0.02	0.33	0.03	1.19	0.04	0.27	0	0	0
385	**Herring**, *raw*	0	0	0	0	0.01	1.25	0.11	1.94	0.01	0.33	0	0	0
386	**Kippers**, *raw*	0	0	0	0	0.02	0.81	0.05	1.58	0.08	0.19	0.03	0	0
387	**Mackerel**, *raw*	0	0	0	0	0	1.01	0.04	1.88	0.03	0.26	0	0	0
388	**Pilchards**, canned in tomato sauce	0	0	0	0	0	0	0.34	0.06	0.87	0.06	0.35	0	0
389	**Salmon**, canned in brine, drained	0	0	0	0	0	0.27	0.04	0.87	0.04	0.17	0.01	0	0
390	**Sardines**, canned in oil, drained	0	0	0	0	0.01	0.44	0.04	1.83	0.05	0.47	0.04	0	0
391	**Sardines**, canned in tomato sauce	0	0	0	0	0.01	0.62	0.05	1.61	0.06	0.41	0.02	0	0
392	**Sprats**, *raw*	0	0	0	0	0	0	0.49	0.06	1.22	0.04	0.33	0	0
393	**Trout, rainbow**, *raw*	0	0	0	0	0	0.19	0.02	0.67	0.01	0.20	0	0	0
394	*grilled*, flesh only	0	0	0	0	0	0.22	0.02	0.66	0.01	0.20	0	0	0
395	**Tuna**, canned in brine, drained	0	0	0	0	0	0.01	Tr	0.15	0.01	0.06	Tr	0	0
396	canned in oil, drained	0	0	0	0	0.01	0.02	0.01	0.79	0.02	0.54	0.04	0	0

Fish and fish products

Monounsaturated fatty acids, g per 100g food

No. Food	10:1	12:1	14:1	cis 15:1	cis 16:1	cis 17:1	18:1	cis/trans 18:1 n-9	18:1 n-7	cis 20:1	cis 22:1	cis/trans 22:1 n-11	22:1 n-9	cis 24:1	trans Monounsatd
White fish															
381 **Cod**, *raw*	0	0	0	0	0.01	0	0.05	N	N	0.01	0	0	0	0	0
382 **Haddock**, *raw*	0	0	0	0	0.02	0	0.06	N	N	0.01	0	0	0	0	Tr
383 **Plaice**, *raw*	0	0	0	0	0.14	0.01	0.18	N	N	0.04	0.02	N	N	0	Tr
Fatty fish															
384 **Eel, jellied**	0	0	0.06	0.02	0.85	0.04	2.43	2.02	0.40	0.10	0	0	0	0	Tr
385 **Herring**, *raw*	0	0	0.01	0.08	1.06	0.07	1.50	0.93	0.15	1.26	2.01	2.01	0	0	Tr
386 **Kippers**, *raw*	0.08	0	0.05	0	0.94	0.11	1.90	1.42	0.32	2.39	3.79	3.63	0.14	0	0
387 **Mackerel**, *raw*	0	0	0	0.03	0.59	0.03	1.91	1.49	0.04	2.10	3.20	3.20	0	0	0
388 **Pilchards**, canned in tomato sauce	0	0	0.02	0.01	0.54	0.07	1.10	0.77	0.31	0.31	0.17	0.13	0.03	0	0
389 **Salmon**, canned in brine, drained	0	0	0.01	0	0.30	0.04	1.14	0.87	0.15	0.85	0.88	0.77	0.09	0.06	0
390 **Sardines**, canned in oil, drained	0	0	0.01	0	0.55	0.06	3.88	N	0.33	0.14	0.11	0	0.11	0	0.09
391 **Sardines**, canned in tomato sauce	0	0	0.02	0.01	0.75	0.07	1.80	1.51	0.29	0.14	0.12	0	0.10	0	0
392 **Sprats**, *raw*	0	0	0.03	0.01	0.72	0.05	1.89	1.56	0.25	0.64	1.33	1.25	0.06	0	0
393 **Trout, rainbow**, *raw*	0	0	0.01	0.02	0.26	0.01	0.98	0.77	0.13	0.29	0.28	0	0.25	0	0
394 *grilled*, flesh only	0	0	0.01	0.02	0.30	0.01	1.08	0.84	0.15	0.32	0.28	0	0.25	0	0
395 **Tuna**, canned in brine, drained	0	0	0	0	0.01	0	0.08	0.07	0.01	0	0	0	0	0	0
396 canned in oil, drained	0	0	0	0	0.02	0.01	2.07	1.92	0.21	0.03	0	0	0	0	0.04

Fish and fish products

Polyunsaturated fatty acids, g per 100g food

No.	Food	cis n-6 18:2	18:3	20:3	20:4	22:4	cis n-3 18:3	18:4	20:5	22:5	22:6	trans Polyunsatd
White fish												
381	**Cod**, *raw*	Tr	0	0	0.02	0	Tr	0	0.08	0.01	0.16	0
382	**Haddock**, *raw*	0.01	0	0	0.01	0.01	Tr	0	0.05	0.01	0.10	Tr
383	**Plaice**, *raw*	0	0	0.01	0	0	0	0.02	0.16	0.04	0.10	Tr
Fatty fish												
384	**Eel, jellied** a	0.15	0	0	0.10	0.03	0.12	0	0.19	0.14	0.15	Tr
385	**Herring**, *raw* b	0.29	0	0	0.04	0	0.18	0.34	0.51	0.11	0.69	Tr
386	**Kippers**, *raw* c	0.18	0.02	0.02	0.03	0	0.24	0.53	1.15	0.10	1.34	0
387	**Mackerel**, *raw* d	0.30	0.04	0.09	0.07	0	0.22	0.64	0.71	0.12	1.10	0
388	**Pilchards**, canned in tomato sauce e	0.12	0.01	0	0.07	0.01	0.07	0.20	1.17	0.23	1.20	0
389	**Salmon**, canned in brine, drained f	0.13	0	0.06	0.11	0.01	0.08	0.22	0.55	0.14	0.86	0
390	**Sardines**, canned in oil, drained g	2.54	0	0	0.04	0	0.36	0.15	0.89	0	0.82	0.03
391	**Sardines**, canned in tomato sauce h	0.96	0.02	0	0.04	0	0.16	0.19	0.89	0.10	0.68	0
392	**Sprats**, *raw* i	0.11	0	0	0.07	0.01	0.08	0.22	0.93	0.10	1.35	0
393	**Trout**, rainbow, *raw* j	0.34	0	0	0.03	0	0.06	0.06	0.23	0.09	0.83	0
394	*grilled*, flesh only k	0.41	0	0	0.08	0	0.07	0.07	0.22	0.07	0.75	0
395	**Tuna**, canned in brine, drained	0.01	0	0	0.02	0	0	0	0.02	0.02	0.14	0
396	canned in oil, drained b	3.17	0	0	0.03	0	0.74	0	0.06	0.04	0.27	0.19

a Contains 0.04g 20:2 per 100g food
b Contains 0.02g 20:2 per 100g food
c Contains 0.01g 16:4, 0.02g 20:2, 0.06 21:5 per 100g food
d Contains 0.03g 20:2 per 100g food
e Contains 0.09g 16:4, 0.04 20:2, 0.05 21:5, 0.02 22:3 per 100g food
f Contains 0.04g 20:2, 0.03g 22:2, 0.03g 22:3 per 100g food

g Contains 0.09g 16:4 per 100g food
h Contains 0.06g 16:4, 0.04g 21:5, 0.03g 22:3 per 100g food
i Contains 0.05g 16:4, 0.02g 20:2, 0.04g 21:5 per 100g food
j Contains 0.01g, 16:4, 0.02g 20:2, 0.01g 22:2 per 100g food
k Contains 0.01g 16:4, 0.02g 20:2, 0.02g 21:5 per 100g food

Fish and fish products

Fat and total fatty acids, g per 100g food

No. Food	Description	Total fat	Satd	cis-Mono unsatd	Polyunsatd Total cis	n-6	n-3	Total trans	Total branched
Crustacea and Molluscs									
397 **Crab**, *boiled*	12 samples, purchased boiled. Light and dark meat	5.5	0.68	1.55	1.36	0.24	1.10	0	0.02
398 **Mussels**, *boiled*	11 fresh and frozen samples	2.7	0.52	0.40	0.76	0.08	0.68	0	0.03
399 **Oysters**, *raw*	Analytical and literature sources	1.3	0.26	0.14	0.41	0.05	0.37	0	0
400 **Prawns**, frozen, *raw*	13 samples, 10 brands	0.6	0.13	0.15	0.13	0.03	0.11	0	Tr
401 **Squid**, *raw*	1 sample	1.7	0.35	0.22	0.61	0.16	0.45	0	0.01
402 **Whelks**, *boiled*	10 samples from assorted outlets	1.2	0.20	0.23	0.28	0.07	0.21	0	0.01
403 **Winkles**, *boiled*	11 samples from assorted outlets	1.2	0.15	0.21	0.35	0.12	0.23	0	0.03
Fish products and dishes									
404 **Fish cakes**, frozen, *raw*	11 packets, 7 brands. Coated in breadcrumbs	3.9	0.59	1.43	0.93	0.60	0.33	0.04	0
405 **Fish fingers**, cod, frozen, *raw*	14 packets, 8 brands. Coated in breadcrumbs	7.8	1.81	1.99	1.38	1.28	0.17	0.27	0
406 **Prawn bhuna**, takeaway	10 samples from different outlets	8.7	1.19	3.42	2.52	1.93	0.58	0.02	0
407 **Prawn madras**, takeaway	10 samples from different outlets	8.3	1.19	3.17	2.44	1.88	0.56	0.02	0
408 **Roe**, cod, hard, *raw*	6 samples from assorted outlets	1.9	0.41	0.37	0.55	0.05	0.49	0	0
409 herring, soft, *raw*	10 samples from assorted outlets	2.6	0.52	0.60	0.70	0.05	0.65	0	0
410 **Scampi**, breaded, *cooked*	2 samples	0.9	0.07	0.34	0.20	0.14	0.06	0	0
411 **Szechuan prawns with vegetables**, takeaway	10 samples from different outlets	4.7	0.56	2.09	1.21	0.90	0.30	0.03	0
412 **Sesame prawn toasts**, takeaway	10 samples from different outlets	29.8	3.29	13.07	8.11	6.52	1.57	0.01	0
413 **Taramasalata**	14 assorted samples. Greek dish based on cod's roe	52.9	3.90	27.71	15.14	10.81	5.00	0.86	0

Saturated fatty acids, g per 100g food

No.	Food	4:0	6:0	8:0	10:0	12:0	14:0	15:0	16:0	17:0	18:0	20:0	22:0	24:0
Crustacea and molluscs														
397	**Crab**, *boiled*	0	0	0	0	0.01	0.10	0.03	0.36	0.02	0.16	0.01	0	0
398	**Mussels**, *boiled*	0	0	0	0	0	0.08	0.01	0.33	0.03	0.05	0.02	0	0
399	**Oysters**, *raw*	0	0	0	0	0	0.04	0.01	0.16	0.01	0.04	0	0	0
400	**Prawns**, frozen, *raw*	0	0	0	0	0	0.02	0	0.08	Tr	0.02	0	0	0
401	**Squid**, *raw*	0	0	0	0	0	0.03	0.01	0.25	0.01	0.04	0	0	0
402	**Whelks**, *boiled*	0	0	0	0	0	0.03	0	0.10	0.01	0.06	0	0	0
403	**Winkles**, *boiled*	0	0	0	0	0	0.03	0	0.08	0.01	0.03	0	0	0
Fish products and dishes														
404	**Fish cakes**, frozen, *raw*	0	0	0	0	0	0.02	Tr	0.44	Tr	0.10	0.01	0	0
405	**Fish fingers**, cod, frozen, *raw*	0	0	0	0	0.01	0.04	0	1.50	0	0.24	0.02	0	0
406	**Prawn bhuna**, takeaway	0	0	0.01	0.01	0.02	0.06	0.01	0.71	0.02	0.24	0.04	0.03	0.04
407	**Prawn madras**, takeaway	0	0	0.01	0.01	0.02	0.06	0.01	0.73	0.02	0.24	0.03	0.05	0
408	**Roe**, cod, hard, *raw*	0	0	0	0	0	0.03	0.01	0.32	0.02	0.04	0	0	0
409	herring, soft, *raw*	0	0	0	0	0	0.04	0.01	0.41	0.01	0.06	0	0	0
410	**Scampi**, breaded, *cooked*	0	0	0	0	0	0	0	0.05	0	0.02	0	0	0
411	**Szechuan prawns with vegetables**, takeaway	0	0	Tr	Tr	Tr	0.01	0.01	0.34	0.01	0.12	0.02	0.01	0.04
412	**Sesame prawn toasts**, takeaway	0	0	Tr	0.01	0.03	0.06	0.01	2.17	0.02	0.72	0.14	0.08	0.05
413	**Taramasalata**	0	0	0	0	0	0.05	0	2.71	0	0.90	0.24	0	0

Fish and fish products

Monounsaturated fatty acids, g per 100g food

No.	Food	cis 10:1	12:1	14:1	15:1	16:1	17:1	18:1	cis/trans 18:1 n-9	cis/trans 18:1 n-7	20:1	cis 22:1	cis/trans 22:1 n-11	cis/trans 22:1 n-9	cis 24:1	trans Monounsatd
Crustacea and molluscs																
397	**Crab**, *boiled*	0	0	0.01	0.02	0.36	0.03	0.86	0.56	0.28	0.22	0.06	0.03	0.01	0	0
398	**Mussels**, *boiled*	0	0	0	0.01	0.21	0.01	0.09	0.06	0	0.08	0	0	0	0	0
399	**Oysters**, *raw*	0	0	0	0	0.03	0	0.05	N	N	0.03	0.03	N	N	0	0
400	**Prawns**, frozen, *raw*	0.01	0	0	0	0.03	0	0.09	0.07	0.03	0.01	0.01	0	0.01	0	0
401	**Squid**, *raw*	0	0	0	0.01	0.01	0	0.13	0.10	0.02	0.06	Tr	0	Tr	0	0
402	**Whelks**, *boiled*	0	0	0	0	0.03	0	0.09	0.05	0.03	0.08	0.03	0	0.02	0	0
403	**Winkles**, *boiled*	0	0	0	0	0	0	0.08	0.05	0.03	0.09	0	0	0	0	0
Fish products and dishes																
404	**Fish cakes**, frozen, *raw*	0	0	0	0	0.02	0	1.32	1.27	0.09	0.05	0.04	0.02	0.02	0	0.03
405	**Fish fingers**, cod, frozen, *raw*	0	0	0	0	0.01	0	1.98	2.01	0.10	0	0	0	0	0	0.20
406	**Prawn bhuna**, takeaway	0	0	0	0	0.04	0.01	3.25	N	N	0.08	0.02	N	N	0	0.02
407	**Prawn madras**, takeaway	0	0	0	0	0.04	0.01	3.01	N	N	0.08	0.02	N	N	0	0.02
408	**Roe**, cod, hard, *raw*	0	0	0	0	0.07	0.01	0.26	0.18	0.06	0.02	0.01	0	0.01	0	0
409	herring, soft, *raw*	0	0	0	0	0.05	0	0.44	0.27	0.14	0.07	0.03	0	0.03	0	0
410	**Scampi**, breaded, *cooked*	0	0	0	0	0	0	0.33	0.33	0.02	0.01	0	0	0	0	0.02
411	**Szechuan prawns with vegetables**, takeaway	0	0	0	0	0.03	0.01	1.98	N	N	0.06	0.01	Tr	Tr	0	0.01
412	**Sesame prawn toasts**, takeaway	0	0	0	0	0.12	0.04	12.54	N	N	0.29	0.08	N	N	0	0.01
413	**Taramasalata**	0	0	0	0	0.14	0	26.57	25.33	1.43	0.71	0.29	0	0.29	0	0.19

Polyunsaturated fatty acids, g per 100g food

No.	Food	cis n-6					18:3	18:4	cis n-3			trans
		18:2	18:3	20:3	20:4	22:4			20:5	22:5	22:6	Polyunsatd
Crustacea and molluscs												
397	**Crab**, boiled [a]	0.02	0.01	0.01	0.12	0.03	0.02	0.05	0.47	0.08	0.45	0
398	**Mussels**, boiled [b]	0.02	0.01	0	0.05	0	0.02	0.05	0.41	0.02	0.16	0
399	**Oysters**, raw	0.02	0	0	0.01	0.02	0.01	0.04	0.14	0.02	0.16	0
400	**Prawns**, frozen, raw	0.01	0	0	0.01	0	0	0	0.06	Tr	0.04	0
401	**Squid**, raw [c]	0.12	0.03	0.01	0.01	0	0	0	0.13	0.01	0.29	0
402	**Whelks**, boiled [d]	0.01	0	0	0.03	0.01	0	0	0.10	0.05	0.05	0
403	**Winkles**, boiled [e]	0.04	0	0.02	0.04	0	0.06	0.03	0.10	0.02	0.00	0
Fish products and dishes												
404	**Fish cakes**, frozen, raw [c]	0.58	0	0	0.01	0	0.18	0	0.05	0.01	0.09	0
405	**Fish fingers**, cod, frozen, raw	1.21	0	0	0	0	0.10	0	0.04	0	0.03	0.08
406	**Prawn bhuna**, takeaway [c]	1.91	0	0	0.02	0	0.53	0	0.03	0.02	0	0
407	**Prawn madras**, takeaway [c]	1.86	0	0	0.02	0	0.48	0.01	0	0.02	0.05	0
408	**Roe**, cod, hard, raw	0.02	0	0	0.04	0	0.01	0	0.16	0.02	0.30	0
409	herring, soft, raw	0.02	0.01	0	0.02	0	0.01	0	0.20	0.05	0.38	0
410	**Scampi**, breaded, cooked	0.14	0	0	0	0	0.06	0	0	0	0	0.01
411	**Szechuan prawns with vegetables**, takeaway [c]	0.88	0	0	0.02	0	0.27	0	0.02	0.01	0	0
412	**Sesame prawn toasts**, takeaway [e]	6.47	0	0	0.03	0.02	1.54	0	0	0	0	0
413	**Taramasalata**	10.52	0	0	0	0	4.43	0	0.14	0.05	0	0.67

a Contains 0.04g 20:2, 0.02g 21:5, 0.12 16 poly, 0.02g 18 poly, 0.04 20 poly, 0.08g 22 poly per 100g food
b Contains 0.01 16:4, 0.01g 21:5, 0.122g 16 poly, 0.03g 20 poly, 0.04g 22 poly per 100g food
c Contains 0.01g 20:2 per 100g food
d Contains 0.02g 16 poly, 0.03g 18 poly, 0.03g 20 poly, 0.04g 22 poly per 100g food
e Contains 0.02g 20:2, 0.03g 16 poly, 0.02g 18 poly, 0.02g 20 poly, 0.03g 22 poly per 100g food

Vegetables and vegetable dishes

Fat and total fatty acids, g per 100g food

No.	Food	Description	Total fat	Satd	cis-Mono unsatd	Polyunsatd Total cis	n-6	n-3	Total trans	Total branched
Potatoes and potato products										
414	**Potato**, *raw*	Different varieties sampled over 2 years Flesh only	0.2	0.04	0	0.12	0.09	0.03	0	0
415	**Potato chips**, fast food chain [a]	Data from TRANSFAIR study; 12 samples from different outlets	11.0	5.96	2.69	0.17	0.16	0.01	0.43	0.08
416	fish and chip shop [b]	Data from TRANSFAIR study; 12 samples from different outlets	8.6	4.58	2.15	0.16	0.18	0.02	0.39	0.06
417	**Microchips**, *microwaved*	10 samples, 3 brands	10.2	3.64	3.18	2.02	2.02	0.05	0.13	0
418	**Potato fritters**, battered, *cooked*	2 samples of different brands	8.5	3.51	2.99	0.94	0.90	0.04	0.02	0.02
419	**Potato waffles**, *baked*	6 samples, 3 brands	12.9	5.01	4.71	1.34	1.34	0.09	0.27	0
Vegetables, beans and lentils										
420	**Ackee**, canned	10 cans, 2 brands, drained	15.2	4.44	7.11	0.60	0.52	0.07	0	0.01
421	**Baked beans**, canned in tomato sauce	10 cans, 7 brands	0.6	0.10	0.06	0.30	0.12	0.18	Tr	0
422	**Chick peas**, hummus	3 samples; tubs and tins	12.6	1.54	4.51	4.95	4.55	0.46	0.07	0
423	**Peas**, *raw*	Whole peas, no pods	1.5	0.56	0.47	0.17	0.13	0.04	Tr	0
Vegetable dishes and products										
424	**Enchiladas**, vegetable, takeaway	8 samples from different outlets	7.1	2.54	2.51	1.02	0.81	0.21	0.20	0
425	**Protein substitute grill/burger**, unbreaded, *baked/grilled*	7 samples, 3 brands	10.5	3.19	1.98	1.85	1.89	0.10	1.38	0
426	**Quorn mycoprotein**, *raw*	10 samples, 2 brands; chunks and mince	3.3	0.55	0.44	1.36	1.02	0.37	0.17	0

a Contains 0.33g unidentified fatty acids per 100g food
b Contains 0.20g unidentified fatty acids per 100g food

Vegetables and vegetable dishes

Saturated fatty acids, g per 100g food

No.	Food	4:0	6:0	8:0	10:0	12:0	14:0	15:0	16:0	17:0	18:0	20:0	22:0	24:0
	Potatoes and potato products													
414	**Potato**, raw	0	0	0	0	0	0	0	0.03	0	0.01	0	0	0
415	**Potato chips**, fast food chain	0	0	0.06	0.03	0.05	0.36	0.11	2.74	0.18	2.42	0.01	0	0
416	fish and chip shop	0	0	0.06	0.02	0.01	0.27	0.09	2.11	0.13	1.87	0.01	0	0
417	**Microchips**, microwaved	0	0	0.01	0	0.02	0.07	0	2.99	0.01	0.47	0.04	0.03	0.02
418	**Potato fritters**, battered, cooked	0	0	0.01	0	0.01	0.08	0	3.05	0.01	0.31	0.02	0.01	0
419	**Potato waffles**, baked	0	0	0.01	0	0.02	0.11	0	4.24	0.01	0.53	0.05	0.01	0.02
	Vegetable, beans and lentils													
420	**Ackee**, canned	0	0	0	0	0	0.01	0	3.02	0.01	1.31	0.07	0.01	0
421	**Baked beans**, canned in tomato sauce	0	0	0	0	0	0	0	0.09	0	0.01	0	0	Tr
422	**Chick peas**, hummus	0	0	0	0	0.01	0.03	0	1.00	0	0.44	0.06	0	0
423	**Peas**, raw	0	0	0	0	0	0.03	0	0.36	0	0.18	0	0	Tr
	Vegetable dishes and products													
424	**Enchiladas**, vegetable, takeaway	0	0.02	0.03	0.07	0.11	0.32	0.04	1.38	0.03	0.50	0.02	0.01	0
425	**Protein substitute grill/burger**, unbreaded, baked/grilled	0	0	0	0	0.34	0.14	0	1.32	0	1.32	0.04	0.04	0
426	**Quorn mycoprotein**, raw	0	0	0	0	0	0.01	0	0.39	0	0.15	0	0	0

Vegetables and vegetable dishes

Monounsaturated fatty acids, g per 100g food

No.	Food	cis 10:1	12:1	14:1	cis 15:1	16:1	17:1	18:1	cis/trans 18:1 n-9	18:1 n-7	20:1	cis 22:1	cis/trans 22:1 n-11	22:1 n-9	cis 24:1	trans Monounsatd
	Potatoes and potato products															
414	**Potato**, raw	0	0	0	0	Tr	0	Tr	0	0	0	0	0	0	0	0
415	**Potato chips**, fast food chain	0	0	0	0	0.16	0	2.52	1.72	0.80	Tr	0	0	0	0	0.42
416	fish and chip shop	0	0	0	0	0.14	0	2.00	1.50	0.41	0.01	0	0	0	0	0.36
417	**Microchips**, microwaved	0	0	0	0	0.02	0.01	3.13	3.06	0	0.02	0	0	0	0	0.07
418	**Potato fritters**, battered, cooked	0	0	0	0	0.01	0	2.96	2.91	0.01	0.01	0.01	0	0.01	0	0.01
419	**Potato waffles**, baked	0	0	0	0	0.02	0	4.64	4.50	0	0.03	0.01	0	0.01	0	0.18
	Vegetables, beans and lentils															
420	**Ackee**, canned	0	0	0	0	0.05	0	7.02	6.92	0.10	0.05	0	0	0	0	0
421	**Baked beans**, canned in tomato sauce	0	0	0	0	0	0	0.06	N	N	0	0	0	0	0	0
422	**Chick peas**, hummus	0	0	0	0	0.02	0	4.45	4.28	0.17	0.04	0	0	0	0	0
423	**Peas**, raw	0	0	0	0	0.03	0	0.44	N	N	0.01	0	0	0	0	0
	Vegetable dishes and products															
424	**Enchiladas**, vegetable, takeaway	0.01	0	0.03	0	0.10	0.02	2.31	N	N	0.04	0.01	0	Tr	0	0.20
425	**Protein substitute grill/burger**, unbreaded, baked/grilled	0	0	0	0	0	0	1.98	1.73	0	0	0	0	0	0	1.25
426	**Quorn mycoprotein**, raw	0	0	0	0	0.02	0.01	0.41	0.41	0	0	0	0	0	0	0.14

Vegetables and vegetable dishes

Polyunsaturated fatty acids, g per 100g food

No.	Food	18:2	*cis* n-6 18:3	20:3	20:4	22:4	*cis* n-3 18:3	18:4	20:5	22:5	22:6	*trans* Polyunsatd
	Potatoes and potato products											
414	**Potato**, *raw*	0.09	0	0	0	0	0.03	0	0	0	0	0
415	**Potato chips**, fast food chain [a]	0.09	0	0	0	0	0.01	0	0	0	0	0.01
416	fish and chip shop [b]	0.09	0	0	0	0	0.02	0	0	0	0	0.04
417	**Microchips**, *microwaved*	1.97	0	0	0	0	0.04	0	0	0	0	0.06
418	**Potato fritters**, battered, *cooked*	0.90	0	0	0	0	0.04	0	0	0	0	0.01
419	**Potato waffles**, *baked*	1.27	0	0	0	0	0.07	0	0	0	0	0.09
	Vegetable, beans and lentils											
420	**Ackee**, canned	0.52	0	0	0	0	0.07	0	0	0	0	0
421	**Baked beans**, canned in tomato sauce	0.12	0	0	0	0	0.18	0	0	0	0	Tr
422	**Chick peas**, hummus	4.55	0	0	0	0	*0.40*	0	0	0	0	0.07
423	**Peas**, *raw*	0.13	0	0	0	0	0.04	0	0	0	0	Tr
	Vegetable dishes and products											
424	**Enchiladas**, vegetable, takeaway	0.78	0.01	0	Tr	0	0.21	0	0	0	0	0.01
425	**Protein substitute grill/burger**, unbreaded, baked/grilled	1.75	0	0	0	0	0.10	0	0	0	0	0.13
426	**Quorn mycoprotein**, *raw* [c]	0.99	0	0	0	0	0.37	0	0	0	0	0.03

[a] Contains 0.07g 16:4 per 100g food
[b] Contains 0.06g 16:4 per 100g food
[c] Contains 0.01g 20 poly per 100g food

Vegetables and vegetable dishes

Fat and total fatty acids, g per 100g food

No.	Food	Description	Total fat	Satd	cis-Mono unsatd	Polyunsatd Total cis	cis n-6	n-3	Total trans	Total branched
Vegetable dishes and products continued										
427	**Vegetables**, *stir-fried*, takeaway	10 samples from different outlets	4.1	0.77	2.17	0.56	0.30	0.23	0.01	0
428	**Vegetable curry**, Thai, takeaway	10 samples from different outlets	8.2	3.64	2.11	1.43	1.15	0.27	0.01	0
429	**Tomatoes**, sun-dried, in olive and sunflower oil	10 samples of different brands bottled in olive and sunflower oils	51.3	6.60	14.64	26.84	26.42	0.72	0.35	0.07
430	**Vegetable balti**, takeaway	10 samples from different outlets	8.1	1.61	3.20	2.22	1.74	0.48	0.03	0
431	**biryani**, takeaway	10 samples from different outlets	7.0	1.47	2.64	1.98	1.59	0.38	0.02	0
432	**Vegeburger mixes**	8 samples, 3 brands	14.7	3.80	3.55	3.47	3.40	0.09	0.94	0
433	**Vegebanger mixes**	8 samples, 2 brands	14.3	4.95	2.55	2.37	2.28	0.13	1.59	0
434	**Vegetable grill/burger**, in crumbs, *baked/grilled*	6 samples, different brands including bean burgers and crispbakes	12.8	2.29	3.53	3.70	3.30	0.58	0.71	0
435	**Vegetable and cheese grill/burger**, in crumbs, *baked/grilled*	9 samples, 6 brands including cheese grills and cheese and onion crispbakes	14.0	4.64	3.68	2.94	2.78	0.32	0.82	0
436	**Vegetable kiev**, *baked*	4 samples, 2 brands including cordon bleu and traditional style	13.7	4.95	3.59	2.50	2.50	0.27	1.89	0
437	**Vegetable pâté**	2 samples; packets and tubs	13.4	4.64	4.67	2.29	1.84	0.54	0.16	0
438	**Vegetable samosa**	4 samples	9.3	1.09	4.65	2.94	2.48	0.77	0.32	0
439	**Vegetarian sausages**, *baked/grilled*	4 samples, 3 brands	9.4	2.33	2.07	1.53	1.65	0.19	2.32	0
Herbs and spices										
440	**Ginger**, ground	8 samples	3.3	1.57	0.95	0.12	0.01	N	0	0
441	**Mustard powder**	4 packets	45.1	2.20	21.90	11.98	7.36	4.62	0	0

Saturated fatty acids, g per 100g food

No.	Food	4:0	6:0	8:0	10:0	12:0	14:0	15:0	16:0	17:0	18:0	20:0	22:0	24:0
	Vegetable dishes and products continued													
427	**Vegetables**, *stir-fried*, takeaway	0	0.02	0.05	Tr	0.01	0.01	0.04	0.40	0.01	0.13	0.03	0.06	0
428	**Vegetable curry**, Thai, takeaway	0	0.02	0.21	0.19	1.61	0.68	0	0.65	0	0.24	0.02	0.02	0.01
429	**Tomatoes**, sun-dried, in olive and sunflower oil	0	0	0	0	0	0	0	4.00	0	2.02	0.18	0.30	0.11
430	**Vegetable balti**, takeaway	0	0.01	0.02	0.03	0.05	0.12	0.02	0.97	0.01	0.29	0.03	0.03	0.01
431	**biryani**, takeaway	0	0	0.02	0.01	0.03	0.08	0.02	0.96	0.01	0.26	0.03	0.03	0.01
432	**Vegeburger mixes**	0	0	0	0	0.01	0.06	0	2.80	0.01	0.83	0.06	0.02	0
433	**Vegebanger mixes**	0	0	0.01	0.01	0.05	0.11	0	3.63	0.01	1.04	0.05	0.02	0.02
434	**Vegetable grill/burger**, in crumbs, *baked/grilled*	0	0	0	0	0.08	0.11	0	1.39	0	0.67	0.05	0	0
435	**Vegetable and cheese grill/burger**, in crumbs, *baked/grilled*	0.10	0.05	0	0.10	0.14	0.44	0	2.64	0	1.16	0	0	0
436	**Vegetable kiev**, *baked*	0.13	0.07	0	0.12	0.08	0.50	0	2.99	0	1.06	0	0	0
437	**Vegetable pâté**	0	0.01	0.19	0.18	2.12	0.95	0	0.91	0	0.25	0.04	0	0
438	**Vegetable samosa**	0	0	0	0	0	0.02	0.01	0.72	0.01	0.24	0.07	0.03	0
439	**Vegetarian sausages**, *baked/grilled*	0	0	0	0	0	0	0	1.76	0	0.58	0	0	0
	Herbs and spices													
440	**Ginger**, ground	0	0	0	0	0.40	0.46	0	0.10	0.50	0.03	0.04	0.05	0
441	**Mustard powder**	0	0	0	0	0	0.04	0	1.41	0	0.43	0.18	0.14	0

Vegetables and vegetable dishes

Monounsaturated fatty acids, g per 100g food

No.	Food	cis							cis/trans		cis		cis/trans		cis	trans
		10:1	12:1	14:1	15:1	16:1	17:1	18:1	18:1 n-9	18:1 n-7	20:1	22:1	22:1 n-11	22:1 n-9	24:1	Monounsatd
	Vegetable dishes continued															
427	**Vegetables**, *stir-fried*, takeaway	0	0.01	0	0	0.01	0	1.12	N	N	0.89	0.14	N	N	0	0.01
428	**Vegetable curry**, Thai, takeaway	0	0	0	0	0.03	0.01	2.01	N	N	0.04	0.01	Tr	Tr	0	0.01
429	**Tomatoes**, sun-dried, in olive and sunflower oil	0	0	0	0	0.11	0	14.42	14.42	0	0.11	0	0	0	0	0
430	**Vegetable balti**, takeaway	0.01	0	0.01	0	0.03	0.01	2.99	N	N	0.11	0.04	N	N	0	0.03
431	**biryani**, takeaway	0	0	0	0	0.03	0.01	2.47	N	N	0.09	0.03	N	N	0	0.02
432	**Vegeburger mixes**	0	0	0	0	0.02	0	3.49	3.34	0.15	0.02	0.01	Tr	Tr	0	0.92
433	**Vegebanger mixes**	0	0	0	0	0.01	0	2.51	2.28	0.23	0.02	0	0	0	0.01	1.56
434	**Vegetable grill/burger**, in crumbs, baked/grilled	0	0	0	0	0	0	3.48	3.19	0.29	0.05	0	0	0	0	0.54
435	**Vegetable and cheese grill/burger**, in crumbs, baked/grilled	0	0	0	0	0.06	0	3.61	3.40	0.21	0	0	0	0	0	0.67
436	**Vegetable kiev**, baked	0	0	0	0	0	0	3.59	3.25	0.34	0	0	0	0	0	1.62
437	**Vegetable pâté**	0	0	0	0	0.07	0.01	4.55	4.41	0.18	0.04	0	0	0	0	0.08
438	**Vegetable samosa**	0	0	0	0	0.02	0.01	4.33	4.05	0.35	0.20	0.10	0	0.10	0	0.07
439	**Vegetarian sausages**, *baked/grilled*	0	0	0	0	0	0	2.07	1.76	0.31	0	0	0	0	0	2.00
	Herbs and spices															
440	**Ginger**, ground	0	0	0	0	0.67	0	0.08	0.05	0.03	0.18	0	0	0	0	0
441	**Mustard powder**	0	0	0	0	0.04	0	7.83	7.83	0	3.86	10.17	0	10.17	0	0

Vegetables and vegetable dishes

Polyunsaturated fatty acids, g per 100g food

No.	Food	cis n-6					cis n-3					trans
		18:2	18:3	20:3	20:4	22:4	18:3	18:4	20:5	22:5	22:6	Polyunsatd
Vegetables dishes and products continued												
427	**Vegetables**, *stir-fried*, takeaway[a]	0.27	0	0.02	*0.01*	0.01	0.06	0	0.04	0.02	0.10	0
428	**Vegetable curry**, Thai, takeaway	1.14	0	0	*0.01*	0	0.26	0	0	0.01	0	0.01
429	**Tomatoes**, sun-dried, in olive and sunflower oil	26.12	0	0	0	0	0.72	0	0	0	0	0.35
430	**Vegetable balti**, takeaway[b]	1.73	0	0	0	0	0.45	0	0	0.01	0.02	0
431	**biryani**, takeaway	1.58	0	0.01	Tr	0	0.34	0	0.01	0.01	0.02	0
432	**Vegeburger mixes**	3.38	0	0	0	0	0.09	0	0	0	0	0.02
433	**Vegebanger mixes**	2.24	0	0	0	0	0.13	0	0	0	0	0.03
434	**Vegetable grill/burger**, in crumbs, *baked/grilled*	3.18	0	0	0	0	0.53	0	0	0	0	0.18
435	**Vegetable and cheese grill/burger**, in crumbs, *baked/grilled*	2.63	0	0	0	0	0.32	0	0	0	0	0.15
436	**Vegetable kiev**, *baked*	2.23	0	0	0	0	0.27	0	0	0	0	0.27
437	**Vegetable pâté**	*1.81*	0	0	0	0	*0.48*	0	0	0	0	0.08
438	**Vegetable samosa**[c]	*2.02*	0	0.18	0	0	*0.57*	0	0	0	0	0.26
439	**Vegetarian sausages**, *baked/grilled*	1.34	0	0	0	0	0.19	0	0	0	0	0.32
Herbs and spices												
440	**Ginger**, ground[d]	Tr	0	0	0.01	0	0	0	0	0	0	0
441	**Mustard powder**[e]	7.00	0	0	0	0	4.62	0	0	0	0	0

a Contains 0.04g 20:2 per 100g food
b Contains 0.01g 20:2 per 100g food
c Contains 0.18g 20:2 per 100g food

d Contains 0.11g 16 poly per 100g food
e Contains 0.25g 20:2, 0.11g 22:3 per 100g food

Fruit, nuts and seeds

Fat and total fatty acids, g per 100g food

No.	Food	Description	Total fat	Satd	cis-Mono unsatd	Polyunsatd Total cis	n-6	n-3	Total trans	Total branched
	Fruit									
442	**Avocado**	10 samples, Fuerte variety	19.3	2.27	14.50	1.68	1.61	0.07	0	0
443	**Banana**	10 samples, flesh only	0.3	0.11	0.04	0.09	0.04	0.05	0	0
444	**Olives**	Mixed sample, bottled in brine, drained	11.0	1.54	7.68	1.23	1.16	0.06	0	0
445	**Yogurt coated nuts and raisins**	Mixed sample	39.5	23.37	8.36	3.70	3.66	0.04	0.11	0
	Nuts and seeds									
446	**Almonds**	10 blanched packets, flaked and ground	55.8	4.43	38.19	10.46	10.19	0.27	0	0
447	**Brazil nuts**	10 packets	68.2	17.41	22.36	25.43	25.43	0	0	0
448	**Coconut**	Mixed samples, fresh	36.0	31.39	2.41	0.62	0.62	0	0	0
449	**Hazelnuts**	10 packets	63.5	4.55	49.23	6.62	6.50	0.12	0	0
450	**Peanuts**, plain	10 packets	46.0	8.66	22.03	13.10	12.75	0.35	0	0
451	**Peanut butter**, smooth	2 samples of the same brand with added vegetable oil	51.8	12.78	19.02	16.84	16.84	0	0.89	0
452	**Poppy seeds**	2 packets	47.1	6.80	10.58	27.65	27.20	0.45	0	0
453	**Pumpkin seeds**	2 packets	45.6	8.59	13.30	21.71	21.58	0.13	0	0
454	**Sesame seeds**	10 packets, with and without hulls	58.0	8.60	22.07	25.50	25.35	0.15	0	0
455	**Shanghai nuts**	4 packets	34.8	10.48	14.01	8.55	8.55	0	0.23	0
456	**Sunflower seeds**	Analytical and literature sourcces	47.5	6.58	10.67	28.15	28.06	0.09	0	0
457	**Walnuts**	10 packets	68.5	7.47	10.67	46.76	39.29	7.47	0	0

Fruits, nuts and seeds

Saturated fatty acids, g per 100g food

No.	Food	4:0	6:0	8:0	10:0	12:0	14:0	15:0	16:0	17:0	18:0	20:0	22:0	24:0
Fruit														
442	**Avocado**	0	0	0	0	0	0	0	2.27	0	0	0	0	0
443	**Banana**	0	0	0	0	0	0	0	0.10	0	0.01	0	0	0
444	**Olives**	0	0	0	0	0	0	0	1.26	0	0.24	0.04	0	0
445	**Yogurt coated nuts and raisins**	0.04	0.04	0.32	0.39	6.01	2.60	0.04	6.86	0.04	6.47	0.25	0.25	0.07
Nuts and seeds														
446	**Almonds**	0	0	0	0	0	0.05	0	3.36	0	0.91	0.11	0	0
447	**Brazil nuts**	0	0	0	0	0	0.13	0	10.17	0	7.11	0	0	0
448	**Coconut**	0	0.24	2.58	2.44	16.42	5.44	0	3.10	0	0.83	0.34	0	0
449	**Hazelnuts**	0	0	0	0	0	0.18	0	3.16	0	1.09	0.12	0	0
450	**Peanuts**, plain	0	0	0	0	0.04	0.22	0	4.71	0	1.19	0.53	1.50	0.48
451	**Peanut butter**, smooth	0	0	0	0	0	0.05	0	6.93	0.05	2.82	0.84	1.54	0.59
452	**Poppy seeds**	0	0	0	0	0	0.05	0	5.22	0.05	1.40	0.09	0	0
453	**Pumpkin seeds**	0	0	0	0	0.04	0	0	6.10	0.04	2.18	0.17	0.04	0
454	**Sesame seeds**	0	0	0	0	0	0.01	0	4.81	0.35	3.12	0.30	0	0
455	**Shanghai nuts**	0	0	0	0	0.10	0.23	0	7.98	0	1.06	0.27	0.57	0.27
456	**Sunflower seeds**	0	0	0	0	0	0.05	0	3.63	0.05	2.18	0.14	0.45	0.09
457	**Walnuts**	0	0	0	0	0	0.72	0	4.91	0	1.38	0.46	0	0

Fruit, nuts and seeds

Monounsaturated fatty acids, g per 100g food

No.	Food	cis						18:1	cis/trans		cis		cis/trans		cis	trans
		10:1	12:1	14:1	15:1	16:1	17:1	18:1	18:1 n-9	18:1 n-7	20:1	22:1	22:1 n-11	22:1 n-9	24:1	Monounsatd
Fruit																
442	**Avocado**	0	0	0	0	0.65	0	13.86	13.86	0	0	0	0	0	0	0
443	**Banana**	0	0	0	0	0.01	0	0.03	0.03	0	0	0	0	0	0	0
444	**Olives**	0	0	0	0	0.11	0	7.57	N	N	0	0	0	0	0	0
445	**Yogurt coated nuts and raisins**	0	0	0	0	0.04	0	8.22	8.25	0.07	0.11	0	0	0	0	0.11
Nuts and seeds																
446	**Almonds**	0	0	0	0	0.37	0	37.82	37.82	0	0	0	0	0	0	0
447	**Brazil nuts**	0	0	0	0	0.26	0	22.10	22.10	0	0	0	0	0	0	0
448	**Coconut**	0	0	0	0	0.14	0	2.27	2.27	0	0	0	0	0	0	0
449	**Hazelnuts**	0	0	0	0	0.18	0	48.99	48.99	0	0.06	0	0	0	0	0
450	**Peanuts**, plain	0	0	0	0	0	0	21.55	21.55	0	0.48	0	0	0	0	0
451	**Peanut butter**, smooth	0	0	0	0	0.05	0	18.52	19.12	0.30	0.45	0	0	0	0	0.89
452	**Poppy seeds**	0	0	0	0	0.09	0.05	10.45	9.73	0.72	0	0	0	0	0	0
453	**Pumpkin seeds**	0	0	0	0	0.04	0	13.25	12.90	0.35	0	0	0	0	0	0
454	**Sesame seeds**	0	0	0	0	0.46	0	21.61	21.16	0.45	0	0	0	0	0	0
455	**Shanghai nuts**	0	0	0	0	0.03	0	13.74	13.77	0.20	0.23	0	0	0	0	0.23
456	**Sunflower seeds**	0	0	0	0	0	0	10.58	10.26	0.32	0.09	0	0	0	0	0
457	**Walnuts**	0	0	0	0	0.13	0	10.54	10.54	0	0	0	0	0	0	0

Fruit, nuts and seeds

Polyunsaturated fatty acids, g per 100g food

No.	Food	cis n-6					cis n-3					trans
		18:2	18:3	20:3	20:4	22:4	18:3	18:4	20:5	22:5	22:6	Polyunsatd
Fruit												
442	**Avocado**	1.53	0	0	0.02	0	0.07	0	0	0	0	0
443	**Banana**	0.04	0	0	0	0	0.05	0	0	0	0	0
444	**Olives**	1.16	0	0	0	0	0.06	0	0	0	0	0
445	**Yogurt coated nuts and raisins**	3.66	0	0	0	0	0.04	0	0	0	0	0
Nuts and seeds												
446	**Almonds**	10.19	0	0	0	0	0.27	0	0	0	0	0
447	**Brazil nuts**	25.43	0	0	0	0	0	0	0	0	0	0
448	**Coconut**	0.62	0	0	0	0	0	0	0	0	0	0
449	**Hazelnuts**	6.50	0	0	0	0	0.12	0	0	0	0	0
450	**Peanuts**, plain	12.75	0	0	0	0	0.35	0	0	0	0	0
451	**Peanut butter**, smooth [a]	16.79	0	0	0	0	0	0	0	0	0	0
452	**Poppy seeds**	27.20	0	0	0	0	0.45	0	0	0	0	0
453	**Pumpkin seeds**	21.58	0	0	0	0	0.13	0	0	0	0	0
454	**Sesame seeds**	25.35	0	0	0	0	0.15	0	0	0	0	0
455	**Shanghai nuts**	8.55	0	0	0	0	0	0	0	0	0	0
456	**Sunflower seeds**	28.06	0	0	0	0	0.09	0	0	0	0	0
457	**Walnuts**	39.29	0	0	0	0	7.47	0	0	0	0	0

[a] Contains 0.05g 20:2 per 100g food

Preserves, confectionery and snacks

Fat and total fatty acids, g per 100g food

No.	Food	Description	Total fat	Satd	cis-Mono unsatd	Polyunsatd Total cis	n-6	n-3	Total trans	Total branched
	Preserves									
458	**Chocolate nut spread**	2 samples of the same brand	33.0	9.34	16.69	5.27	5.27	0.13	0.25	0
459	**Lemon curd**	4 samples, 4 brands	4.9	1.52	1.75	1.07	0.94	0.15	0.35	0
460	**Marzipan**	10 samples, white and yellow	12.7	1.03	7.99	3.13	3.12	0.01	0	0
	Chocolate confectionery									
461	**Bounty bar**	8 samples; plain and milk chocolate	26.1	21.18	3.07	0.40	0.37	0.02	0.12	0
462	**Chocolate covered caramels**	18 samples, 5 brands	21.7	10.75	6.66	0.66	0.66	0.04	2.50	0.04
463	**Chocolate bar with wafer/biscuit and fruit**	28 samples of Lion Bar, Picnic, Ballisto, Crispy Caramel	27.4	13.36	8.54	2.20	2.15	0.05	1.94	0.13
464	**Chocolate**, cooking	10 samples	34.8	28.01	3.76	0.67	0.63	0.03	0.80	0
465	couverture	3 samples, 2 brands; milk and plain	37.4	22.60	11.66	1.47	1.36	0.11	0	0
466	fancy and filled	10 samples of different brands	21.3	11.24	6.86	0.99	0.95	0.04	1.09	0.04
467	milk	12 bars, 5 brands including Dairy Milk, Galaxy, chocolate buttons	30.7	18.06	9.45	1.11	1.02	0.09	0.38	0.12
468	plain	6 bars, 3 brands	28.0	16.76	8.91	1.04	0.96	0.08	0.05	0
469	**Creme eggs**	10 samples	14.6	8.72	4.54	0.49	0.45	0.04	0.21	0
470	**Kit Kat**	2 samples	26.0	16.18	7.41	0.65	0.64	0.05	0.17	0.09
471	**Mars bar**	8 samples	18.9	9.79	5.21	0.70	0.70	0.09	2.03	0.13
472	**Milky way**	10 samples	15.8	8.44	4.83	0.51	0.54	0.05	1.21	0.03
473	**Twix**	10 samples	24.5	10.28	6.59	0.80	0.97	0.07	6.33	0
474	**Wispa bar**	4 samples	34.2	19.16	11.44	1.67	1.54	0.13	0.36	0.10

Preserves, confectionery and snacks

Saturated fatty acids, g per 100g food

No.	Food	4:0	6:0	8:0	10:0	12:0	14:0	15:0	16:0	17:0	18:0	20:0	22:0	24:0
	Preserves													
458	**Chocolate nut spread**	0	0	0	0	0.13	0.35	0.03	6.59	0.03	2.05	0.13	0.03	0
459	**Lemon curd**	0	0	0	0	0	0.07	0	1.06	0.01	0.32	0.03	0.02	0
460	**Marzipan**	0	0	0	0	0	0.01	0	0.79	0.01	0.18	0.01	0.01	0.01
	Chocolate confectionery													
461	**Bounty bar**	0.07	0.17	1.75	1.27	8.51	2.92	0.02	3.29	0.05	2.99	0.07	0.05	0
462	**Chocolate covered caramels**	0.06	0.04	0.04	0.06	0.25	0.37	0.04	4.64	0.04	4.99	0.17	0.04	0
463	**Chocolate bar with wafer/biscuit and fruit**	0.08	0.05	0.13	0.18	1.55	0.86	0.05	5.47	0.05	4.64	0.16	0.13	0
464	**Chocolate**, cooking	N	0.07	1.00	0.90	12.21	4.19	N	3.93	0.03	5.49	0.13	0.03	0.03
465	couverture	0.14	0.07	0.04	0.11	0.14	0.46	N	9.62	0.11	11.48	0.32	0.07	0.04
466	fancy and filled	0.08	0.06	0.08	0.10	0.38	0.45	0.04	4.76	0.04	5.08	0.14	0.02	0
467	milk	0.15	0.09	0.06	0.15	0.18	0.64	0.09	8.26	0.09	8.11	0.23	0.03	0
468	plain	0.03	0.03	0.03	0.03	0.05	0.19	0.03	7.28	0.05	8.78	0.21	0.05	0
469	**Creme eggs**	N	0.04	0.03	0.08	0.10	0.33	0.07	3.95	0.04	3.91	0.13	0.03	0.01
470	**Kit Kat**	0.07	0.04	0.14	0.20	2.21	1.20	0.06	5.78	0.07	6.41	0.00	0	0
471	**Mars bar**	0.22	0.13	0.09	0.16	0.22	0.52	0.07	4.19	0.05	3.99	0.13	0.04	0
472	**Milky way**	0.03	0.03	0.05	0.06	0.17	0.27	0.03	3.46	0.05	4.06	0.15	0.09	0
473	**Twix**	0.07	0.05	0.05	0.10	0.17	0.34	0.05	4.66	0.07	4.54	0.14	0.05	0
474	**Wispa bar**	0.20	0.13	0.10	0.20	0.29	1.05	0.16	5.85	0.16	10.85	0.10	0.07	0

Preserves, confectionery and snacks

Monounsaturated fatty acids, g per 100g food

No.	Food	10:1	12:1	14:1	cis 15:1	16:1	17:1	18:1	cis/trans 18:1 n-9	cis/trans 18:1 n-7	cis 20:1	cis 22:1	cis/trans 22:1 n-11	cis/trans 22:1 n-9	cis 24:1	trans Monounsatd
	Preserves															
458	**Chocolate nut spread**	0	0	0	0	0.09	0	16.50	15.21	0.28	0.06	0.03	0.03	0	0	0.13
459	**Lemon curd**	0	0	0	0	0.03	0.01	1.65	1.51	0	0.03	0.01	Tr	Tr	0	0.27
460	**Marzipan**	0	0	0	0	0.06	0.01	7.88	7.72	0	0.01	0.01	Tr	Tr	0.01	0
	Chocolate confectionery															
461	**Bounty bar**	0	0	0	0	0.05	0	3.02	3.04	0.10	0	0	0	0	0	0.12
462	**Chocolate covered caramels**	0	0	0.02	0	0.06	0	6.56	7.49	0.87	0.02	0	0	0	0	2.45
463	**Chocolate bar with wafer/biscuit and fruit**	0	0	0.03	0	0.10	0	8.38	7.33	0.55	0.03	0	0	0	0	1.94
464	**Chocolate**, cooking	0	0	0	0	0.03	0	3.69	3.56	0	0	0.03	N	N	0	0.80
465	couverture	0	0	0	0	0.14	0.04	11.44	11.44	0	0.04	0	0	0	0	0
466	fancy and filled	0	0	0.02	0	0.06	0.02	6.76	7.41	0.28	0	0	0	0	0	1.09
467	milk	0.03	0	0.06	0	0.12	0.03	9.22	9.25	0.23	0	0	0	0	0	0.38
468	plain	0	0	0.03	0	0.08	0	8.81	8.78	0.08	0	0	0	0	0	0.05
469	**Creme eggs**	0	0	0	0	0.07	0.01	4.44	4.33	N	0.01	0	0	0	0	0.20
470	**Kit Kat**	0.01	0	0.03	0	0.11	0.01	7.26	N	N	0	0	0	0	0	0.13
471	**Mars bar**	0.04	0	0.05	0	0.07	0.02	4.99	5.69	0.79	0.04	0	0	0	0	1.94
472	**Milky way**	0.03	0	0.02	0	0.06	0.02	4.70	5.05	0.53	0.02	0	0	0	0	1.13
473	**Twix**	0	0	0.02	0	0.07	0.02	6.45	11.08	1.18	0.02	0	0	0	0	6.08
474	**Wispa bar**	0.02	0	0.07	0.02	0.26	0.03	10.60	10.92	0.03	0.46	0	0	0	0	0.36

Polyunsaturated fatty acids, g per 100g food

No.	Food	18:2	18:3	*cis* n-6 20:3	20:4	22:4	18:3	18:4	*cis* n-3 20:5	22:5	22:6	*trans* Polyunsatd
Preserves												
458	**Chocolate nut spread**	*5.14*	0	0	0	0	0.13	0	0	0	0	0.13
459	**Lemon curd**	0.91	0	0	0	0	0.15	0	0	0	0	0.08
460	**Marzipan**	3.12	0	0	0	0	0.01	0	0	0	0	0
Chocolate confectionery												
461	**Bounty bar**	0.37	0	0	0	0	0.02	0	0	0	0	0
462	**Chocolate covered caramels**	*0.62*	0	0	0	0	0.04	0	0	0	0	0.04
463	**Chocolate bar with wafer/biscuit and fruit**	2.15	0	0	0	0	0.05	0	0	0	0	0
464	**Chocolate**, cooking	0.63	0	0	0	0	0.03	0	0	0	0	0
465	couverture	1.36	0	0	0	0	0.11	0	0	0	0	0
466	fancy and filled	0.95	0	0	0	0	0.04	0	0	0	0	0
467	milk	1.02	0	0	0	0	0.09	0	0	0	0	0
468	plain	0.96	0	0	0	0	0.08	0	0	0	0	0
469	**Creme eggs**	0.45	0	0	0	0	0.04	0	0	0	0	0
470	**Kit Kat**	*0.61*	0	0	0	0	0.05	0	0	0	0	0.04
471	**Mars bar**	*0.61*	0	0	0	0	0.09	0	0	0	0	0.09
472	**Milky way**	*0.47*	0	0	0	0	0.05	0	0	0	0	0.08
473	**Twix**	*0.72*	0	0	0	0	0.07	0	0	0	0	0.24
474	**Wispa bar** [a]	1.50	0	0	0	0	0.13	0	0	0	0	0

[a] Contains 0.03g 20:2 per 100g food

Preserves, confectionery and snacks

Fat and total fatty acids, g per 100g food

No.	Food	Description	Total fat	Satd	cis-Mono unsatd	Polyunsatd Total cis	n-6	n-3	Total trans	Total branched
Non chocolate confectionery										
475	**Cereal chewy bar**	17 bars of different brands; assorted types	16.4	5.03	5.61	1.72	1.80	0.03	3.16	0
476	**Cereal crunchy bar**	12 bars of different brands; assorted types	22.2	4.48	10.08	5.35	4.67	0.74	1.32	0
477	**Chew sweets**	7 samples of different brands including Starburst, Chewitts and Fruitella	5.6	2.99	0.98	0.10	0.17	0	1.28	0
478	**Fudge**	4 samples, 2 brands	16.8	14.15	1.00	0.18	0.16	0.05	0.58	0.08
479	**Liquorice**	2 samples	1.3	0.39	0.34	0.47	0.42	0.06	0.04	0
480	**Liquorice allsorts**	4 samples of different brands including mini allsorts	5.2	3.61	0.65	0.16	0.15	0.02	0.56	0
481	**Toffees**, mixed	13 samples, 4 brands including cream and plain varieties	18.6	9.44	4.46	0.39	0.55	0.05	3.31	0.12
Savoury snacks										
482	**Bombay mix**	20 packets, savoury mix made from gram flour, assorted peas, lentils, nuts and seeds	32.9	3.94	16.27	11.00	9.72	1.55	0.20	0.01
483	**Breadsticks**	10 samples, brands	8.4	5.85	1.30	0.88	0.84	0.04	0	0
484	**Corn snacks**	20 samples, 7 brands including Wotsits, Monster Munch and Nik Naks	31.9	11.78	12.78	5.79	5.11	0.71	0.15	0.01
485	**Corn and starch snacks**	10 samples, 5 brands including Skip	31.6	13.75	12.63	3.63	3.66	0.12	0.21	0
486	**Maize and rice flour snacks**	20 samples of Frazzles and Bacon Streaks	20.9	6.33	9.25	4.26	3.54	0.84	0.16	0
487	**Mixed cereal and potato snacks**	10 samples of Ringos	23.1	7.31	7.42	7.04	6.56	0.60	0.31	0
488	**Pork scratchings**	23 samples, 4 brands	46.0	16.58	23.75	3.47	3.21	0.26	0.18	0
489	**Poppadums**, takeaway	10 samples from different outlets	38.8	7.95	16.42	12.52	10.19	2.33	0.07	0

Preserves, confectionery and snacks

Saturated fatty acids, g per 100g food

No.	Food	4:0	6:0	8:0	10:0	12:0	14:0	15:0	16:0	17:0	18:0	20:0	22:0	24:0
Non chocolate confectionery														
475	**Cereal chewy bar**	0	0	0.02	0.02	0.16	0.11	0.01	2.73	0.03	1.95	0	0	0
476	**Cereal crunchy bar**	0	0.02	0.11	0.08	0.74	0.30	0	2.08	0.02	0.83	0.11	0.13	0.06
477	**Chew sweets**	0	0.01	0.12	0.10	0.99	0.37	0	0.70	0.01	0.65	0.02	0.01	0
478	**Fudge**	0.10	0.10	0.55	0.56	5.96	2.25	0.05	2.07	0.03	2.44	0.05	0	0
479	**Liquorice**	0	0	0	0	0.05	0.02	0	0.21	0	0.09	0.01	0	0
480	**Liquorice allsorts**	0	0.02	0.29	0.23	1.74	0.62	0	0.43	0	0.24	0.01	0	0
481	**Toffees**, mixed	0.14	0.11	0.16	0.21	1.51	1.08	0.07	4.13	0.05	1.88	0.07	0	0
Savoury snacks														
482	**Bombay mix**	0	0	0	0.01	0.08	0.05	0.01	2.69	0.02	0.80	0.27	0	0
483	**Breadsticks**	0	0.03	0.38	0.30	2.33	0.92	0	1.65	0	0.22	0.01	0	0
484	**Corn snacks**	0	0	0	0.04	0.11	0.39	0.02	9.86	0.03	1.20	0.11	0	0
485	**Corn and starch snacks**	0	0	0.03	0	0.06	0.30	0	11.78	0.06	1.33	0.12	0.03	0.03
486	**Maize and rice flour snacks**	0	0	0.02	0	0.04	0.12	0	5.19	0.02	0.78	0.08	0.04	0.04
487	**Mixed cereal and potato snacks**	0	0	0	0	0.02	0.13	0	5.98	0.02	0.95	0.09	0.07	0.04
488	**Pork scratchings**	0	0	0	0.04	0.04	0.75	0	10.99	0.09	4.57	0.09	0	0
489	**Poppadums**, takeaway	0	0	0.01	Tr	0.06	0.14	0.04	6.13	0.03	1.16	0.18	0.14	0.04

Preserves, confectionery and snacks

Monounsaturated fatty acids, g per 100g food

No.	Food	cis 10:1	12:1	14:1	15:1	16:1	17:1	18:1	cis/trans 18:1 n-9	18:1 n-7	cis 20:1	22:1	cis/trans 22:1 n-11	22:1 n-9	cis 24:1	trans Monounsatd
Non chocolate confectionery																
475	Cereal chewy bar	0	0	0	0	0.03	0.01	5.57	N	N	0	0	0	0	0	3.04
476	Cereal crunchy bar	0	0	0	0	0.04	0.02	9.74	7.39	3.40	0.19	0.08	0.02	0	0	1.25
477	Chew sweets	0	0.02	0	0	0.01	0	0.94	1.99	0.11	0.02	0	0	0	0	1.22
478	Fudge	0.02	0	0.03	0	0.05	0	0.88	1.16	0.21	0.02	0	0	0	0	0.55
479	Liquorice	0	0	0	0	0.02	Tr	0.31	0.30	0.02	0.01	0.01	0	Tr	0	0.03
480	Liquorice allsorts	0	0.01	0	0	Tr	0	0.63	0.51	0.62	Tr	0	0	0	0	0.55
481	Toffees, mixed	0.02	0	0.05	0	0.07	0.02	4.25	6.17	0.71	0.04	0.02	0	0.02	0	3.09
Savoury snacks																
482	Bombay mix	0	0	0	0.01	0.06	0.13	15.82	1.74	13.44	0.27	0	0	0	0	0.06
483	Breadsticks	0	0	0	0	0.01	0	1.28	1.26	0.02	0.01	0	0	0	0	0
484	Corn snacks	0	0	0	0	0.08	0.01	12.59	12.26	0.43	0.10	0	0	0	0	0.11
485	Corn and starch snacks	0	0	0	0	0.06	0	12.51	12.30	0	0.06	0	0	0	0	0.06
486	Maize and rice flour snacks	0	0	0	0	0.04	0.02	9.03	8.67	0	0.12	0.02	N	N	0.02	0.04
487	Mixed cereal and potato snacks	0	0	0	0	0.02	0	7.33	7.09	0	0.04	0.02	N	N	0	0.09
488	Pork scratchings	0	0	0	0	1.80	0.13	21.46	17.72	2.11	0.35	0	0	0	0	0.18
489	Poppadums, takeaway	0.02	0	0	0	0.08	0.01	15.83	N	N	0.38	0.11	N	N	0	0.07

Preserves, confectionery and snacks

Polyunsaturated fatty acids, g per 100g food

No.	Food	cis n-6					cis n-3					trans
		18:2	18:3	20:3	20:4	22:4	18:3	18:4	20:5	22:5	22:6	Polyunsatd
Non chocolate confectionery												
475	**Cereal chewy bar**	1.69	0	0	0	0	0.03	0	0	0	0	0.12
476	**Cereal crunchy bar**	4.67	0	0	0	0	0.68	0	0	0	0	0.06
477	**Chew sweets** [a]	0.09	0	0	0.01	0	0	0	0	0	0	0.07
478	**Fudge**	0.13	0	0	0	0	0.05	0	0	0	0	0.03
479	**Liquorice**	0.41	0	0	0	0	0.06	0	0	0	0	0.01
480	**Liquorice allsorts**	0.15	0	0	0	0	0.01	0	0	0	0	0
481	**Toffees**, mixed	0.34	0	0	0	0	0.05	0	0	0	0	0.21
Savoury snacks												
482	**Bombay mix** [b]	9.57	0	0	0	0	1.41	0	0	0	0	0.14
483	**Breadsticks**	0.84	0	0	0	0	0.04	0	0	0	0	0
484	**Corn snacks**	5.11	0	0	0	0	0.67	0	0	0	0	0.04
485	**Corn and starch snacks**	3.53	0	0	0	0	0.09	0	0	0	0	0.15
486	**Maize and rice flour snacks**	3.50	0	0	0	0	0.76	0	0	0	0	0.12
487	**Mixed cereal and potato snacks**	6.45	0	0	0	0	0.60	0	0	0	0	0
488	**Pork scratchings** [c]	2.99	0	0.09	0	0	0.26	0	0	0	0	0.22
489	**Poppadums**, takeaway	10.19	0	0	0	0	2.28	0	0	0.05	0	0

a Contains 0.01g 20:2 per 100g food
b Contains 0.02g 20:2 per 100g food

c Contains 0.13g 20:2 per 100g food

Preserves, confectionery and snacks

Fat and total fatty acids, g per 100g food

Savoury snacks continued

No.	Food	Description	Total fat	Satd	cis-Mono unsatd	Polyunsatd Total cis	n-6	n-3	Total trans	Total branched
490	**Potato crisps**, plain	20 samples, 8 brands; mixed plain and flavoured	34.2	14.04	13.51	4.97	N	N	N	0
491	reduced fat	20 samples of different brands; mixed plain and flavoured	21.5	9.34	8.64	2.50	2.40	0.12	0.07	0
492	**Potato and corn sticks**	20 samples, 9 brands including mixed plain and flavoured, including chipsticks and crunchy sticks	23.6	6.61	10.36	5.30	4.73	0.74	0.29	0
493	**Potato and tapioca snacks**	20 samples, 3 brands; assorted types including Waffles, Bitza Pizza, Wickettes	24.8	5.43	10.57	7.47	7.23	0.43	0.26	0
494	**Potato rings**	10 samples, 3 brands; assorted flavours; Hula Hoop type	32.0	13.92	12.60	3.82	3.85	0.15	0.24	0
495	**Prawn crackers**, takeaway	10 samples from different outlets	39.0	3.62	22.35	10.98	7.87	3.14	0.12	0
496	**Punjabi puri**	20 samples, 3 brands; assorted flavours	35.1	8.13	13.86	11.40	11.22	0.18	0.12	0
497	**Puffed potato products**	20 samples, 3 brands; assorted flavours including Quavers, Snaps and Chinese style crackers	33.0	9.75	15.58	4.39	4.13	0.91	1.92	0
498	**Tortilla chips**	20 samples, 6 brands, maize chips	22.6	3.98	6.91	6.31	6.87	0.13	4.41	0
499	**Twiglets**	20 samples; savoury wholewheat snacks	11.7	4.94	4.01	1.80	N	N	N	0
500	**Wheat crunchies**	20 samples, 3 brands; assorted flavours	20.7	9.06	8.00	2.48	2.48	0.07	0.20	0

Saturated fatty acids, g per 100g food

No.	Food	4:0	6:0	8:0	10:0	12:0	14:0	15:0	16:0	17:0	18:0	20:0	22:0	24:0
	Savoury snacks continued													
490	**Potato crisps**, plain	0	0	0	0.01	0.09	0.35	0.03	12.05	0.03	1.36	0.12	0	0
491	reduced fat	0	0	0	0.01	0.07	0.23	0.01	8.05	0.02	0.88	0.07	0	0
492	**Potato and corn sticks**	0	0	0	0	0.05	0.11	0	5.14	0.02	0.83	0.14	0.20	0.11
493	**Potato and tapioca snacks**	0	0	0.02	0.02	0.02	0.07	0	3.39	0.02	0.71	0.24	0.57	0.36
494	**Potato rings**	0	0	0	0	0.06	0.31	0	11.90	0.03	1.44	0.12	0.03	0.03
495	**Prawn crackers**, takeaway	0	0	0.01	0.01	0.01	0.03	0	2.29	0.02	0.76	0.25	0.22	0.03
496	**Punjabi puri**	0	0	0	0.30	0.64	1.36	0.11	4.45	0.03	0.98	0.27	0	0
497	**Puffed potato products**	0	0	0.03	0	0.06	0.22	0	7.95	0.03	1.14	0.16	0.09	0.06
498	**Tortilla chips**	0	0	0	0	0	0.09	0	2.68	0	1.06	0.09	0.06	0
499	**Twiglets**	0	0	0	0.01	0.03	0.14	0.01	4.09	0.01	0.62	0.02	0	0
500	**Wheat crunchies**	0	0	0	0.01	0.06	0.21	0.01	7.87	0.02	0.81	0.07	0	0

Preserves, confectionery and snacks

Monounsaturated fatty acids, g per 100g food

No.	Food	cis							cis/trans		cis		cis/trans		cis	trans
		10:1	12:1	14:1	15:1	16:1	17:1	18:1	18:1 n-9	18:1 n-7	20:1	22:1	22:1 n-11	22:1 n-9	24:1	Monounsatd
	Savoury snacks continued															
490	**Potato crisps**, plain	0	0	0	0	0.08	0.01	13.37	13.11	0.39	0.06	0	0	0	0	0.18
491	reduced fat	0	0	0	0	0.05	0.01	8.54	8.43	0.18	0.05	0	0	0	0	0.06
492	**Potato and corn sticks**	0	0	0	0	0.05	0.02	10.06	9.75	0.32	0.18	0.02	N	N	0.02	0.11
493	**Potato and tapioca snacks**	0	0	0	0	0.02	0.02	10.12	9.91	0.21	0.33	0.05	N	N	0.02	0.07
494	**Potato rings**	0	0	0	0	0.06	0	12.45	12.24	0.21	0.09	0	0	0	0	0.06
495	**Prawn crackers**, takeaway	0.01	0	0	0	0.10	0.06	21.50	N	N	0.55	0.13	N	N	0	0.06
496	**Punjabi puri**	0	0	0.10	0	0.10	0.01	13.47	13.29	0.23	0.17	0	0	0	0	0.05
497	**Puffed potato products**	0	0	0	0	0.06	0.03	15.17	14.07	1.10	0.25	0.03	N	N	0.03	1.26
498	**Tortilla chips**	0	0	0	0	0	0	6.81	9.61	0.76	0.09	0.02	0	0.02	0	3.72
499	**Twiglets**	0	0	0	0	0.03	0	3.97	3.55	0.29	0.01	0	0	0	0	0.43
500	**Wheat crunchies**	0	0	0	0	0.04	0	7.92	7.86	0.18	0.05	0	0	0	0	0.13

Preserves, confectionery and snacks

Polyunsaturated fatty acids, g per 100g food

No.	Food	cis n-6 18:2	18:3	20:3	20:4	22:4	18:3	18:4	cis n-3 20:5	22:5	22:6	trans Polyunsatd
	Savoury snacks continued											
490	**Potato crisps**, plain	4.65	0	0	0	0	0.32	0	0	0	0	N
491	reduced fat	2.40	0	0	0	0	0.11	0	0	0	0	0.01
492	**Potato and corn sticks**	4.65	0	0	0	0	0.65	0	0	0	0	0.18
493	**Potato and tapioca snacks**	7.11	0	0	0	0	0.36	0	0	0	0	0.19
494	**Potato rings**	3.70	0	0	0	0	0.12	0	0	0	0	0.18
495	**Prawn crackers**, takeaway [a]	7.81	0	0	0	0.01	2.96	0	0	0.10	0.08	0.05
496	**Punjabi puri**	11.22	0	0	0	0	0.18	0	0	0	0	0
497	**Puffed potato products**	3.69	0	0	0	0	0.69	0	0	0	0	0.66
498	**Tortilla chips**	6.18	0	0	0	0	0.13	0	0	0	0	0.69
499	**Twiglets**	1.72	0	0	0	0	0.08	0	0	0	0	N
500	**Wheat crunchies**	2.41	0	0	0	0	0.06	0	0	0	0	0.07

[a] Contains 0.03g 20:2 per 100g food

Beverages and soups

Fat and total fatty acids, g per 100g food

No.	Food	Description	Total fat	Satd	cis-Mono unsatd	Polyunsatd Total cis	n-6	n-3	Total trans	Total branched
Beverages										
501	**Cream liqueur**, whisky based [a]	Data from TRANSFAIR study; 2 samples of Baileys	15.5	9.78	3.30	0.25	0.19	0.07	0.52	0.05
502	**Drinking chocolate**	10 samples, 6 brands	5.8	3.41	1.81	0.30	0.28	*0.02*	0	0
503	**Instant drinks powder**, chocolate, low calorie	10 samples, 4 brands	11.1	8.05	1.56	0.85	0.80	*0.05*	0.09	0
504	malted	10 samples, 3 brands	9.5	8.69	0.18	0.06	0.06	0	0.06	0
Soups and stock										
505	**Canned cream of chicken soup**, *as served*	12 samples, 5 brands	2.9	0.42	1.51	0.73	0.56	0.23	0.10	0
506	**Canned cream of mushroom soup**, *as served*	10 samples, 3 brands	3.0	0.31	1.67	0.80	0.60	0.27	0.08	0
507	**Canned cream of tomato soup**, *as served*	10 samples, 6 brands	3.0	0.45	1.50	0.72	0.56	0.25	0.18	0
508	**Dried cream of tomato soup**	10 packets, 4 brands including cream of tomato	14.3	6.80	3.41	0.33	0.50	0.05	3.09	0
509	**Dried instant soup**	9 samples, 5 brands, not tomato	14.3	6.94	3.58	0.38	0.48	0.05	2.78	0
510	**Stock cubes**, beef	10 samples, 6 brands including Bovril, Oxo and own brands	9.2	3.51	3.15	1.26	1.14	0.25	0.26	0.13

[a] Contains 0.59g unidentified fatty acids per 100g food

Saturated fatty acids, g per 100g food

No.	Food	4:0	6:0	8:0	10:0	12:0	14:0	15:0	16:0	17:0	18:0	20:0	22:0	24:0
Beverages														
501	**Cream liqueur**, whisky based	0	0	0.24	0.41	0.53	1.69	0.39	4.68	0.15	1.67	0.02	0.01	0
502	**Drinking chocolate**	0	0	0	0	0	0	0	1.39	0.01	1.94	0.06	0.01	0
503	**Instant drinks powder**, chocolate, low calorie	0	0.09	0.77	0.38	2.32	0.74	0	1.65	0	2.03	0.05	0	0
504	malted	0	0.02	0.38	0.35	3.65	1.11	0	0.90	0	2.24	0.03	0	0
Soups and stock														
505	**Canned cream of chicken soup**, *as served*	0.01	Tr	Tr	Tr	0.01	0.02	0	0.27	Tr	0.09	0.01	0.01	Tr
506	**Canned cream of mushroom soup**, *as served*	0	Tr	Tr	Tr	0.01	0.02	0	0.17	Tr	0.06	0.02	0.01	0.01
507	**Canned cream of tomato soup**, *as served*	0.01	0.01	Tr	0.01	0.01	0.03	0.01	0.24	Tr	0.09	0.02	0.01	Tr
508	**Dried cream of tomato soup**	0	0	0.01	0.01	0.04	0.15	0	5.02	Tr	1.43	0.07	0.03	0.01
509	**Dried instant soup**	0	0	0.01	0.01	0.11	0.16	0	4.95	0.03	1.56	0.07	0.03	0.01
510	**Stock cubes**, beef	0	0.07	0.05	0.10	0.13	0.42	0.05	1.87	0.05	0.74	0.03	0.01	0.01

No.	Food	cis							cis/trans		cis	cis	cis/trans		cis	trans
		10:1	12:1	14:1	15:1	16:1	17:1	18:1	18:1 n-9	18:1 n-7	20:1	22:1	22:1 n-11	22:1 n-9	24:1	Monounsatd
Beverages																
501	**Cream liqueur**, whisky based	0	0	0	0	0.22	0	3.08	N	N	0	0	0	0	0	0.49
502	**Drinking chocolate**	0	0	0	0	0.02	0	1.80	1.80	0	0	0	0	0	0	0
503	**Instant drinks powder**, chocolate, low calorie	0	0	0	0	0	0	1.56	1.56	0	0	0	0	0	0	0.09
504	malted	0	0	0	0	0	0	0.18	0.18	0	0	0	0	0	0	0.06
Soups and stock																
505	**Canned cream of chicken soup**, as served	0	0	0	0	0.02	0.01	1.42	1.33	0.10	0.04	0.01	Tr	Tr	0.01	0.04
506	**Canned cream of mushroom soup**, as served	0	0	0	0	0.01	0	1.60	1.48	0.11	0.04	0.01	Tr	Tr	0.01	0.01
507	**Canned cream of tomato soup**, as served	0	0	0	0	0.01	Tr	1.44	1.35	0.09	0.03	0.01	Tr	Tr	0.01	0.10
508	**Dried cream of tomato soup**	0	0	0	0	0.01	0	3.37	2.87	0.50	0.01	0.01	Tr	Tr	Tr	2.85
509	**Dried instant soup**	0	0	0	0	0.03	0	3.53	3.05	0.48	0.01	0	0	0	0.01	2.62
510	**Stock cubes**, beef	0	0	0	0	0.14	0.03	2.92	2.92	0	0.05	0	0	0	0.01	0.13

Beverages and soups

Polyunsaturated fatty acids, g per 100g food

No.	Food	18:2	18:3	cis n-6 20:3	20:4	22:4	18:3	18:4	cis n-3 20:5	22:5	22:6	trans Polyunsatd
	Beverages											
501	**Cream liqueur**, whisky based	0.16	0	0.01	0.01	0	0.06	0	0.01	0	0	0.03
502	**Drinking chocolate**	0.28	0	0	0	0	0.02	0	0	0	0	0
503	**Instant drinks powder**, chocolate, low calorie	0.80	0	0	0	0	0.05	0	0	0	0	0
504	malted [a]	0.06	0	0	0	0	0	0	0	0	0	0
	Soups and stock											
505	**Canned cream of chicken soup**, *as served*	0.55	0	0	0	0	0.18	0	0	0	0	0.06
506	**Canned cream of mushroom soup**, *as served*	0.59	0	0	0	0	0.22	0	0	0	0	0.08
507	**Canned cream of tomato soup**, *as served*	0.54	0	0	0	0	0.18	0	0	0	0	0.08
508	**Dried cream of tomato soup**	0.29	0	0	0	0	0.04	0	0	0	0	0.23
509	**Dried instant soup**	0.34	0	0	0	0	0.04	0	0	0	0	0.15
510	**Stock cubes**, beef [b]	1.04	0	0	0	0	0.21	0	0	0	0	0.13

[a] Contains 0.01g 20 poly per 100g food
[b] Contains 0.06g 20 poly, 0.01g 22 poly per 100g food

Fat and total fatty acids, g per 100g food

No.	Food	Description	Total fat	Satd	cis-Mono unsatd	Polyunsatd Total cis	n-6	n-3	Total trans	Total branched
	Sauces and dressings									
511	**Cook-in sauces**	2 cans of the same brand	2.9	0.29	1.58	0.90	0.64	0.27	0	0
512	**Dressing**, blue cheese	2 samples of different brands	46.3	8.50	9.69	24.08	21.25	4.38	1.99	0
513	French	4 samples, 2 brands	49.4	7.98	10.59	28.17	27.35	1.07	0.26	0.05
514	reduced calorie	4 samples, 2 brands including thousand island	11.2	1.59	3.27	5.52	5.09	0.49	0.06	0.07
515	thousand island	2 samples of different brands	30.2	3.75	11.69	12.76	10.57	2.66	0.46	0
516	tofu	2 samples of different brands	30.7	5.05	8.19	15.38	13.30	2.67	0.73	0
517	**Horseradish sauce**	8 samples, 5 brands; creamed and plain	8.4	1.09	3.73	2.99	2.43	0.66	0.15	0.01
518	**Mayonnaise**	3 samples, 2 brands	75.6	11.42	17.86	41.46	40.22	2.15	1.23	0
519	reduced calorie	3 samples, 2 brands	28.1	4.19	6.75	15.36	14.90	0.79	0.47	0
520	**Salad cream**	3 samples, 2 brands	31.0	3.29	11.44	14.44	13.55	1.00	0.10	0.07
521	reduced calorie	2 samples of different brands	17.2	2.55	4.69	9.14	8.86	0.28	0	0
522	**Sandwich spread**	10 samples, 4 brands	9.8	1.12	3.48	4.68	4.44	0.34	0.11	0

Saturated fatty acids, g per 100g food

No.	Food	4:0	6:0	8:0	10:0	12:0	14:0	15:0	16:0	17:0	18:0	20:0	22:0	24:0
	Sauces and dressings													
511	**Cook-in sauces**	0.03	0.01	0.01	0.01	0.01	0.01	0	0.18	0	0.04	0	0	0
512	**Dressing**, blue cheese	0.09	0.04	0	0	0.09	0.22	0.04	5.53	0.04	2.12	0.13	0.18	0
513	French	0	0	0	0.02	0.05	0.23	0.03	4.99	0.06	2.60	0	0	0
514	reduced calorie	0	0	0	0	0	0.02	0	1.12	0.01	0.43	0	0	0
515	thousand island	0	0	0	0	0	0.03	0	2.66	0.03	0.90	0.14	0	0
516	tofu	0	0	0	0	0	0.03	0	3.08	0.03	1.44	0.12	0.29	0.06
517	**Horseradish sauce**	0	0	0	0	0	0.01	0.01	0.75	0.01	0.23	0.05	0.03	0
518	**Mayonnaise**	0	0	0	0	0	0.08	0.02	7.73	0.07	3.52	0	0	0
519	reduced calorie	0	0	0	0	0	0.03	0	2.87	0.05	1.24	0	0	0
520	**Salad cream**	0	0	0	0	0	0.04	0	2.15	0	1.10	0	0	0
521	reduced calorie	0	0	0	0	0	0.05	0	1.64	0	0.74	0.05	0.07	0
522	**Sandwich spread**	0	0	0	0	0	0.01	0	0.62	0.01	0.36	0.04	0.07	0.03

Monounsaturated fatty acids, g per 100g food

No.	Food	cis 10:1	12:1	14:1	15:1	16:1	17:1	18:1	cis/trans 18:1 n-9	18:1 n-7	cis 20:1	cis 22:1	cis/trans 22:1 n-11	22:1 n-9	cis 24:1	trans Monounsatd
	Sauces and dressings															
511	**Cook-in sauces**	0	0	0	0	0	0	1.56	1.47	0.08	0.02	0	0	0	0	0
512	**Dressing**, blue cheese	0	0	0.04	0	0.09	0	9.47	9.30	0.62	0.09	0	0	0	0	0.44
513	French	0	0	0	0	0.08	0	10.51	9.99	0.52	0	0	0	0	0	0
514	reduced calorie	0	0	0	0	0.02	0.01	3.25	3.08	0.16	0	0	0	0	0	0
515	thousand island	0	0	0	0	0.09	0	11.26	10.60	0.72	0.26	0.09	N	N	0	0.06
516	tofu	0	0	0	0	0.03	0.03	7.98	7.75	0.38	0.15	0	0	0	0	0.15
517	**Horseradish sauce**	0	0	0	0	0.04	0	3.38	3.12	0.27	0.15	0.15	0	0	0	0.04
518	**Mayonnaise**	0	0	0	0	0.10	0.04	17.72	17.14	0.90	0	0	0	0	0	0.33
519	reduced calorie	0	0	0	0	0.06	0	6.69	6.48	0.34	0	0	0	0	0	0.13
520	**Salad cream**	0	0	0	0	0.07	0	11.37	10.80	0.57	0	0	0	0	0	0
521	reduced calorie	0	0	0	0	0.03	0	3.88	3.58	0.28	0.20	0.58	0.58	0	0	0
522	**Sandwich spread**	0	0	0	0	0.02	0.01	3.31	3.16	0	0.07	0.07	0.07	0	0.01	0.02

Sauces and dressings

Polyunsaturated fatty acids, g per 100g food

511 to 522

No.	Food	cis n-6					18:3	18:4	cis n-3			trans
		18:2	18:3	20:3	20:4	22:4			20:5	22:5	22:6	Polyunsatd
	Sauces and dressings											
511	**Cook-in sauces**	0.64	0	0	0	0	0.27	0	0	0	0	0
512	**Dressing**, blue cheese	20.54	0	0	0	0	3.54	0	0	0	0	1.55
513	French	27.25	0	0	0	0	0.92	0	0	0	0	0.26
514	reduced calorie	5.07	0	0	0	0	0.45	0	0	0	0	0.06
515	thousand island	10.36	0	0	0	0	2.40	0	0	0	0	0.40
516	tofu [a]	12.83	0	0	0	0	2.38	0	0	0	0	0.59
517	**Horseradish sauce** [b]	2.39	0	0	0.12	0	0.57	0	0	0.01	0.02	0.11
518	**Mayonnaise**	39.51	0	0	0	0	1.95	0	0	0	0	0.90
519	reduced calorie	14.65	0	0	0	0	0.70	0	0	0	0	0.34
520	**Salad cream**	13.48	0	0	0	0	0.95	0	0	0	0	0.10
521	reduced calorie	8.86	0	0	0	0	0.28	0	0	0	0	0
522	**Sandwich spread**	4.38	0	0	0	0	0.30	0	0	0	0	0.09

a Contains 0.06g 20:2 per 100g food
b Contains 0.01g 20:2 per 100g food

The
Appendices

FATTY ACID NOMENCLATURE

Most fatty acids in food consist of a straight chain of carbon atoms ending with a carboxyl group. There may be double bonds between some of these carbon atoms. Saturated fatty acids contain no double bonds; monounsaturated fatty acids contain one double bond, and polyunsaturated fatty acids contain two or more double bonds. The main column headings on the second, third and fourth pages for each food in this supplement indicate the number of carbon atoms and the number of double bonds in each fatty acid, so that, for example, 4:0 is a fatty acid with four carbon atoms and no double bonds, 10:1 contains ten carbon atoms and one double bond, and 18:2 contains eighteen carbon atoms and two double bonds.

The position of the double bond along the chain, starting from the methyl or non carboxyl end of the chain, is physiologically important and may be indicated by the letter n. This is most important for polyunsaturated fatty acids, where there are two main series with the first double bond starting either three or six carbon atoms from the methyl end. These are designated n-3 or n-6, and are shown separately on page 4 for each food. The main positional isomers of 18:1 and 22:1 have also been indicated on page 3 for each food. In some literature the position is indicated by the prefix ω (omega) instead of n, and a less used alternative system indicates the position of the double bonds starting from the carboxyl end with a Δ (delta).

A further factor of physiological significance is the configuration of double bonds, which may be either cis or trans. In most naturally occurring mono- and polyunsaturated fatty acids, they are in the cis form, but, particularly in fats that have been hardened industrially or in ruminants, some may be trans. The amounts of each have been indicated separately for a number of the fatty acids in this supplement.

The following diagram shows the position and configuration of the double bonds in linoleic acid (cis 18:2 n-6).

Many of the fatty acids also have common names, which are given in the table below.

NOMENCLATURE OF INDIVIDUAL FATTY ACIDS

Names of fatty acids occurring in the tables:

No of carbon atoms and double bonds	Systematic name	Common name	Fatty acid family
Saturated acids			
4:0	Butanoic acid	Butyric acid	
6:0	Hexanoic acid	Caproic acid	
8:0	Octanoic acid	Caprylic acid	
10.0	Decanoic acid	Capric acid	
11:0			
12:0	Dodecanoic acid	Lauric acid	
13:0			
14:0	Tetradecanoic acid	Myristic acid	
15:0	Pentadecanoic acid		
16:0	Hexadecanoic acid	Palmitic acid	
17:0	Heptadecanoic acid	Margaric acid	
18:0	Octadecanoic acid	Stearic acid	
20:0	Eicosanoic acid	Arachidic acid	
		Arachic acid	
22:0	Docosanoic acid	Behenic acid	
24:0	Tetracosanoic acid	Lignoceric acid	
Monounsaturated acids			
10:1	Decenoic acid		
12:1	Dodecenoic acid		
14:1	Tetradecenoic acid	Myristoleic acid	
15:1	Pentadecenoic acid		
16:1	Hexadecenoic acid	Palmitoleic acid	
17:1	Heptadecenoic acid		
18:1 (*cis*)	Octadecenoic acid	Oleic acid	n-9
		cis Vaccenic acid	n-7
18:1 (*trans*)		Elaidic acid	n-9
		trans Vaccenic acid	n-7
20:1	Eicosenoic acid	Eicosenic acid	
		Gadoleic acid	
22:1	Docosenoic acid	Cetoleic acid	n-11
		Erucic acid	n-9
24:1	Tetracosenoic acid	Nervonic acid	
		Selacholeic acid	

No of carbon atoms and double bonds	Systematic name	Common name	Fatty acid family
Polyunsaturated acids			
16.2	Hexadecadienoic acid		
16:4	Hexadecatetraenoic acid		
18:2	Octadecadienoic acid	Linoleic acid	n-6
18.3	Octadecatrienoic acid	γ-linolenic acid	n-6
		α-linolenic acid	n-3
18:4	Octadecatetraenoic acid	Stearidonic acid	n-3
20:2	Eicosadienoic acid		
20:3	Eicosatrienoic acid	dihomo-γ-linolenic acid	n-6
20:4	Eicosatetraenoic acid	Arachidonic acid	n-6
20:5	Eicosapentaenoic acids	EPA	n-3
21:5	Heneicosapentaenoic acid		
22:2	Docosadienoic acid		
22:3	Docosatrienoic acid		
22:4	Docosatetraenoic acid	Adrenic acid	n-6
22:5	Decosapentaenoic acid	Clupanodonic acid	n-3
22:6	Decosahexaenoic acid	DHA	n-3
18 poly	Unidentified C18 fatty acids		
20 poly	Unidentified C20 fatty acids		
22 poly	Unidentified C22 fatty acids		

CHOLESTEROL

Cholesterol values for selected foods that have not been included in previous supplements, or revisions of previous values.

No.	FOOD	CHOLESTEROL (mg/100g of food)
Rolls		
17	**Croissants**, savoury, retail	10.1
18	sweet, retail	40.3
Biscuits		
31	**Cheese sandwich biscuits**	14.9
34	**Chocolate chip cookies**	1.3
Cakes		
55	**Chocolate cup cake**	11.6
56	**Chocolate covered marshmallow teacake**	7.0
60	**Gâteau, chocolate based**	55.7
61	fruit, frozen	52.5
65	**Torte**, frozen/chilled, fruit	42.0
Puddings		
76	**Rice desserts**, with fruit, individual, chilled	2.4
Savouries		
79	**Cheese nachos**, takeaway	32.8
80	**Pizza**, thin base, cheese and tomato, takeaway	22.0
81	-, fish topped, takeaway	25.4
82	deep pan, cheese and tomato, takeaway	13.2
83	-, meat topped, takeaway	16.9
Cows' milk		
84	**Skimmed milk**, pasteurised, average	3.7
85	**Semi-skimmed milk**, pasteurised, average	5.9
86	**Whole milk**, pasteurised, average	13.5
Other milks		
98	**Goats milk**	11.0
102	**Sheeps milk**	12.0

No.	FOOD	CHOLESTEROL (mg/100g of food)
Yogurts		
137	**Whole milk yogurt**, fruit	3.4
138	infant	3.1
139	**Low fat yogurt**, plain	1.3
140	hazelnut	1.6
143	**Virtually fat free/diet yogurt**, fruit, twin pot	1.2
144	**Fromage frais**, plain	8.6
145	fruit, children's	3.3
146	**Greek style yogurt**, plain	16.7
Ice creams		
149	**Ice cream**, reduced calorie	12.0
Puddings and chilled desserts		
151	**Cheesecake**, fruit, individual	14.7
152	**Chocolate dairy dessert**, individual	21.1
154	**Custard**, ready to eat	2.1
155	**Fruit fool**, individual	10.3
158	**Trifle**, chocolate, individual	14.6
159	fruit	12.8
Eggs		
160	**Chicken eggs**	380.5
161	**Duck eggs**	678.5
162	**Quail eggs**	900.3
Savoury egg dishes		
163	**Egg fried rice**, takeaway	16.3
164	**Scotch egg**, retail	164.8
Spreading fats		
171	**Butter**, spreadable	282.0
175	**Margarine**, catering	70.0
178	soya	1.4
Offal		
286	**Kidney**, lamb, *grilled*	452.0
291	**Tripe**, dressed, *stewed*	64.1
Meat pies and pastries		
317	**Sausage roll**, flaky pastry, *cooked*	60.9
318	**Spring rolls**, meat, takeaway	7.4

No.	FOOD	CHOLESTEROL (mg/100g of food)

Sausages
331	**Turkey sausages**, *raw*	88.9

Meat dishes
360	**Beef stir-fried with green peppers in black bean** sauce, *takeaway*	24.9
362	**Chicken chop suey**, takeaway	31.9
363	**Chicken fried rice**, takeaway	31.3
368	**Chicken with cashew nuts**, takeaway	29.5
379	**Sweet and sour pork**, battered, takeaway	27.4

Fish products and dishes
405	**Fish fingers**, cod, frozen, *raw*	24.7
406	**Prawn bhuna**, takeaway	158.0
407	**Prawn madras**, takeaway	130.2
410	**Scampi**, breaded, *cooked*	8.5
411	**Szechuan prawns with vegetables**, takeaway	55.6
412	**Sesame prawn toasts**, takeaway	43.7

Vegetable products and dishes
424	**Enchiladas**, vegetable, takeaway	12.6
425	**Protein substitute grill/burger**, unbreaded, *baked/grilled*	15.3
427	**Vegetables**, *stir-fried*, takeaway	1.2
428	**Vegetable curry**, Thai, takeaway	19.5
430	**Vegetable balti**, takeaway	5.1
434	**Vegetable grill/burger**, in crumbs, *baked/grilled*	5.1
436	**Vegetable kiev**, *baked*	12.8
438	**Vegetable samosa**	0.8

Herbs and spices
441	**Mustard powder**	4.1

Fruit nuts and seeds
445	**Yogurt coated nuts and raisins**	9.5
455	**Shanghai nuts**	0.3

Savoury snacks
488	**Pork scratchings**	129.2
489	**Poppadums**, takeaway	1.9
495	**Prawn crackers**, takeaway	0.3
498	**Tortilla chips**	0.6

Beverages
503	**Instant drinks powder**, chocolate, low calorie	2.6
504	malted	5.4

PHYTOSTEROLS

Plants contain a number of phytosterols (plant sterols) which are distinct from cholesterol. In plant oils, the three most common sterols are β-sitosterol, campesterol and stigmasterol. There may also be measurable amounts of at least nine other phytosterols.

The amounts of the five main phytosterols are shown below for selected foods.

Phytosterols, mg per 100g food

No.	FOOD	Brassicca-sterol	Campe-sterol	Stigma-sterol	β-Sito-sterol	5-Avena-sterol	Other	Total phytosterols
Rolls								
17	**Croissants**, savoury, retail	1.9	15.5	2.1	24.4	0.9	0.7	45.6
18	sweet, retail	0.4	1.6	0	9.3	0.4	0	11.7
Biscuits								
31	**Cheese sandwich biscuits**	0.3	6.7	2.0	21.9	0.9	0.6	32.4
34	**Chocolate chip cookies**	0	6.3	3.3	20.4	1.0	0	31.0
Cakes								
55	**Chocolate cup cake**	0.4	2.7	0.4	4.2	0.3	0	8.0
56	**Chocolate covered marshmallow teacake**	0	6.0	6.0	21.0	0	0	33.0
60	**Gâteau, chocolate based**	0	2.7	1.0	7.1	0	0	10.8
61	fruit, frozen	0	2.4	0	5.9	0	0	8.3
65	**Torte**, frozen/chilled, fruit	0	2.5	0	6.1	0	0	8.6
Savouries								
79	**Cheese nachos**, takeaway	3.6	15.5	2.7	29.1	1.2	2.8	54.9
80	**Pizza**, thin base, cheese and tomato, takeaway	0	7.9	0	16.8	0	0	24.7
81	-, fish topped, takeaway	0	6.3	0	14.2	0	0	20.5
82	deep pan, cheese and tomato, takeaway	1.0	8.7	0	19.3	0	0	29.0
83	-, meat topped, takeaway	1.0	7.3	0	14.9	0	0	23.2
Cows' milk								
84	**Skimmed milk**, pasteurised, average	0	0	0	0.2	0	0.2	0.4
85	**Semi-skimmed milk**, pasteurised, average	0.1	0	0	0	0	0.1	0.2
86	**Whole milk**, pasteurised, average	0.1	0.1	0	0.1	0	0.1	0.5
92	**Coffee compliment powder**	0	1.0	2.0	9.0	0	0	12.0
Yogurts								
140	**Low fat yogurt**, hazelnut	0	0	0	1.3	0	0	1.3
145	**Fromage frais**, fruit, children's	0.2	0.2	0	0	0	0	0.4
147	**Soya, alternative to yogurt**, fruit	0	0.1	0	0.2	0	0	0.3

No	FOOD	Brassica-sterol	Campe-sterol	Stigma-sterol	β-Sito-sterol	5-Avena-sterol	Other	Total phytosterols
Puddings and chilled desserts								
151	**Cheesecake**, fruit, individual	0	0.4	0	1.1	0	0	1.5
152	**Chocolate dairy dessert**, individual	0.1	1.1	2.0	4.8	0.2	0.1	8.3
157	**Mousse**, chocolate, reduced fat, individual	0	0.1	0.3	0.4	0	0	0.8
158	**Trifle**, chocolate, individual	0	0.4	0.3	1.2	0	0	1.9
159	fruit	0	0.2	0	0.5	0	0	0.7
Savoury egg dishes								
163	**Egg fried rice**, takeaway	1.8	6.8	0.3	19.3	0	0.5	28.7
164	**Scotch egg**, retail	1.3	6.6	0.7	12.1	0.2	0	20.8
Cooking fats								
167	**Ghee**, vegetable	Tr	11.0	7.0	33.0	0	0	51.0
170	**Suet**, vegetable	0	30.2	8.0	72.2	0	6.4	116.8
Spreading fats								
175	**Margarine**, catering	27.0	101.0	9.0	151.0	11.0	0	299.0
178	soya	0	49.1	39.6	121.7	2.7	4.6	217.7
182	**Fat spread**, 60% fat, with olive oil	Tr	64.0	12.0	147.0	0	0	223.0
Beef								
212	**Fillet steak**, *cooked from steakhouse*, lean only	0.3	0.8	0	0.4	0	0	1.5
214	**Minced beef**, *raw*	0	0.4	0	0	0	0	0.4
215	*stewed*	0	0.2	0	0.1	0	0	0.3
216	extra lean, *raw*	0	0.4	0	0	0	0	0.4
217	**Rump steak**, *fried in corn oil*, lean only	0	1.9	0.4	5.6	0.2	0.2	8.3
219	**Sirloin steak**, *grilled medium-rare*, lean only	0.1	0.3	0	0	0	0	0.4
221	**Topside**, *roasted medium-rare*, lean and fat	0.1	0.4	0	0	0	0	0.5
Veal								
222	**Veal escalope**, *fried in corn oil*, lean only	0.1	4.1	1.2	12.7	0.6	0.9	19.6
Lamb								
232	**Loin chops**, *grilled*, lean only	0.1	0.5	0	0.3	0	0	0.9
233	**Minced lamb**, *stewed*, lean and fat	0.2	0.6	0	0.2	0	0	1.0
Pork								
243	**Belly joint/slices**, *roasted*, lean and fat	0	0.8	0	0.2	0	0	1.0
246	**Diced pork**, *casseroled*, lean only	0	2.6	0.6	5.8	0.3	0.4	9.7
247	**Fillet strips**, *stir-fried in corn oil*, lean only	0	6.7	2.1	19.1	1.1	1.4	30.4
249	**Loin chops**, *grilled*, lean and fat	0.1	0.6	0	0.2	0	0	0.9
252	**Loin steaks**, *fried in corn oil*, lean only	0.1	1.8	0.4	4.0	0.2	0.4	6.9
255	**Minced pork**, *raw*, lean and fat	0	0.5	0	0.1	0	0	0.6
256	*dry fried and stewed*, lean and fat	0	0.8	0	0.3	0	0	1.1

No. FOOD	Brassicca-sterol	Campe-sterol	Stigma-sterol	β-Sito-sterol	5-Avena-sterol	Other	Total phytosterols
Poultry							
263 **Chicken**, breast, *grilled*, meat only	0.3	0.9	0	0.3	0.1	0	1.6
269 **Turkey**, skin, *raw*	0.8	0.9	0	0.3	0.1	0	2.1
270 dark meat, *roasted*	0.6	0.9	0	0.2	0.1	0	1.8
271 light meat, *roasted*	0.2	0.5	0	0.1	0	0	0.8
272 breast fillet, *grilled*, meat only	0.2	0.4	0	0.1	0	0	0.7
274 thighs, diced, *casseroled*	0.3	2.0	0.3	3.3	0.3	0.2	6.4
Other poultry							
275 **Duck**, *raw*, meat only	0	2.0	0	1.0	0	0	3.0
277 *roasted*, meat only	0	1.0	0	0	0	0	1.0
Game							
279 **Pheasant**, *casseroled*, meat only	0	3.0	1.0	7.0	0	0	11.0
281 **Rabbit**, *stewed*, meat only	0	0	0	1.0	0	0	1.0
282 **Venison**, *raw*	0	0	1.0	0	0	0	1.0
283 *casseroled*, meat only	0	0	1.0	1.0	1.0	0	3.0
Offal							
291 **Tripe**, dressed, *stewed*	0	0.4	0	0.1	0	0	0.5
Bacon and ham							
292 **Bacon rashers**, back, *raw*	0	0	0	0.2	0	0	0.2
293 *grilled*	0	0	0	0.1	0	0	0.1
294 **middle**, *raw*	0	0.5	0	0.3	0	0	0.8
295 -, *fried in corn oil*	0.4	6.4	1.4	17.0	0.9	1.0	27.1
296 -, *grilled*	0	0	0	0.1	0	0	0.1
297 **streaky**, *raw*	0	0.5	0	0.2	0	0	0.7
298 -, *grilled*	0	0.7	0	0.1	0	0	0.8
299 **Bacon loin steaks**, *grilled*	0	0.6	0	0.1	0	0	0.7
300 **Ham**, gammon joint, *boiled*	0	0.7	0	0.1	0	0	0.8
301 **gammon rashers**, *grilled*	0	0.7	0	0.1	0	0	0.8
Beefburgers							
310 **Economy burgers**, *grilled*	0.1	1.5	0	3.1	0.1	0.2	5.0
Meat pies and pastries							
312 **Beef pie**, chilled/frozen, *baked*	1.1	8.2	0.9	17.1	0.7	0.4	28.4
313 **Chicken pie**, individual, chilled/frozen, *baked*	1.1	8.6	1.1	17.6	0.6	0.3	29.3
314 **Cornish pastie**, *cooked*	0	7.0	2.0	15.7	0	0	24.8
316 **Lamb samosa**, retail	2.9	13.6	2.2	23.5	1.2	2.4	45.8
317 **Sausage roll**, flaky pastry, *cooked*	0	8.4	0	21.8	0	0	30.2
318 **Spring rolls**, meat, takeaway	3.6	16.0	2.4	25.4	0.6	2.2	50.2
320 **Steak and kidney pudding**, canned	0	5.3	0	8.6	0	0	14.0

No. FOOD	Brassicca-sterol	Campe-sterol	Stigma-sterol	β-Sito-sterol	5-Avena-sterol	Other	Total phytosterols
Continental style sausages							
335 **Saveloy**, unbattered, takeaway	0	2.0	1.0	7.0	0	0	10
Other meat products							
337 **Chicken in crumbs**, stuffed with cheese and vegetables, chilled/frozen, *baked*	0.3	5.8	3.2	13.6	0.5	1.6	25.0
338 **Chicken kiev**, frozen, *baked*	0	5.0	3.0	12.0	0	0	20
341 **Chicken tikka**, chilled, *reheated*	0.6	3.6	0.9	5.4	0.1	0.9	11.5
343 **Doner kebabs**, *cooked*, lean only	0	0.6	0	0.6	0	0	1.2
344 **Faggots in gravy**	0.3	2.7	0.2	5.4	0.3	0.3	9.2
345 **Ham and pork**, chopped, canned	0.1	0.5	0	0.1	0	0.1	0.8
354 **Rissoles**, savoury, *cooked*	0	2.2	0	3.2	0	0	5.3
355 **Shish kebabs**, *cooked*, lean only	0.6	3.2	0	3.3	0	0.3	7.4
Meat dishes							
358 **Beef stew**, meat only	0	0.6	0.3	2.2	0.1	0.2	3.4
360 **Beef stir-fried with green peppers in black bean sauce**, takeaway	3.2	13.0	1.0	19.7	0.6	1.8	39.3
361 **Cannelloni**, chilled/frozen, *reheated*	0	0	0	4.5	0	0	4.5
362 **Chicken chop suey**, takeaway	2.9	12.2	3.1	21.1	1.2	2.1	42.6
363 **Chicken fried rice**, takeaway	2.6	9.7	0.3	14.7	0.3	1.8	29.4
368 **Chicken with cashew nuts**, takeaway	3.0	12.8	1.5	26.8	1.8	2.5	48.4
369 **Cottage/Shepherd's pie**, chilled/frozen, *reheated*	0.1	1.0	0.2	2.3	0.2	0.2	4.0
371 **Irish stew**, canned	0	0	0	1.5	0	0	1.5
373 **Lasagne**, chilled/frozen, *reheated*	0.2	2.0	1.1	5.2	0.3	0.2	8.9
374 **Moussaka**, chilled/frozen/longlife, *reheated*	2.0	8.6	0.5	12.8	0.8	0.3	25.0
379 **Sweet and sour pork**, battered, takeaway	6.2	25.5	1.5	38.0	1.5	3.4	76.1
380 **Tagliatelle with ham**, mushrooms and cheese, chilled/frozen/longlife, *reheated*	0.2	3.8	0.2	3.8	0.4	0.8	9.2
Fish products and dishes							
405 **Fish fingers**, cod, frozen, *raw*	0	3.2	2.4	11.2	2.4	0	19.3
406 **Prawn bhuna**, takeaway	8.7	30.4	4.0	50	8.8	3.5	105.4
407 **Prawn madras**, takeaway	6.3	23.5	3.5	37.3	1.4	2.3	74.3
410 **Scampi**, breaded, *cooked*	0	0	0	1.8	0	0	1.8
411 **Szechuan prawns with vegetables**, takeaway	2.1	8.4	1.1	14.2	0.8	1.4	28.0
412 **Sesame prawn toasts**, takeaway	12.5	51.4	4.7	85.4	6.5	7.7	168.2
413 **Taramasalata**	34.5	149.3	1.8	190.4	14.3	4.2	394.5
Potato products							
417 **Microchips**, *microwaved*	0	2.0	1.0	8.0	1.0	0	12.0
419 **Potato waffles**, *baked*	3.0	4.0	2.0	10	0	0	19.0

No. FOOD	Brassicca-sterol	Campe-sterol	Stigma-sterol	β-Sito-sterol	5-Avena-sterol	Other	Total phytosterols
Vegetables, beans and lentils							
420 **Ackee**, canned	0	0.3	12.0	4.8	0	0.8	17.9
422 **Chick peas**, hummus	1.3	10.3	3.0	31.3	2.5	0	48.4
Vegetable dishes and products							
424 Enchiladas, vegetable, takeaway	1.2	6.9	4.0	18.6	0.5	1.5	32.7
425 **Protein substitute grill/burger**, unbreaded, *baked/grilled*	0	6.4	2.7	21.7	1.0	2.9	34.7
426 **Quorn mycoprotein**, *raw*	0.3	1.5	0	3.8	3.5	63.9	73.0
427 **Vegetables**, *stir-fried*, takeaway	2.9	12.8	2.5	13.7	0.3	1.4	33.6
428 **Vegetable curry**, Thai, takeaway	2.6	12.6	3.2	27.6	1.2	1.9	49.1
430 **Vegetable balti**, takeaway	3.6	14.8	2.2	26.6	0.6	2.4	50.2
432 **Vegeburger mixes**	0	15.0	5.0	51.0	2.0	0	73.0
433 **Vegebanger mixes**	0	16.0	3.0	45.0	0	0	64.0
434 **Vegetable grill/burger**, in crumbs, *baked/grilled*	*1.9*	*13.8*	*4.4*	*29.7*	*0*	*1.9*	*51.7*
436 **Vegetable kiev**, *baked*	0	6.3	2.6	18.9	0	0	27.8
437 **Vegetable pâté**	2.9	22.5	1.5	23.9	0	0	50.7
438 **Vegetable samosa**	7.0	30	0.3	48.3	0	0	85.6
439 **Vegetarian sausages**, *baked/grilled*	0	4.8	3.1	12.1	0	0.2	20.2
Herbs and spices							
440 **Ginger**, ground	0	4.8	5.4	23.2	1.7	0.7	35.6
441 **Mustard powder**	23.5	71.3	0	137.1	14.9	0	246.7
Fruit							
445 **Yogurt coated nuts and raisins**	0	4.3	2.8	19.4	2.4	0	28.8
Nuts and seeds							
452 **Poppy seeds**	0	19.3	2.8	57.9	8.9	0	89.0
453 **Pumpkin seeds**	0	0	0	48.7	0	40.1	88.8
455 **Shanghai nuts**	0	7.0	4.2	28.5	7.0	0.7	47.3
456 **Sunflower seeds**	0	20.9	15.8	140.4	23.8	21.6	222.5
Preserves							
458 **Chocolate nut spread**	0	4.5	2.9	21.8	1.0	0.3	30.4
459 **Lemon curd**	1.0	4.0	1.0	6.0	0	0	12.0
460 **Marzipan**	0	0	0	29.0	0	0	29.0
Chocolate confectionery							
464 **Chocolate**, cooking	0	5.0	6.0	27.0	0	0	38.0
465 couverture	0	49.0	11.0	132.0	0	0	192.0
469 **Creme eggs**	0	3.0	6.0	14.0	0	0	23.0
474 **Wispa bar**	0	9.6	5.7	36.86	0	0	52.2

Phytosterols, mg per 100g food

No. FOOD	Brassicca-sterol	Campe-sterol	Stigma-sterol	β-Sito-sterol	5-Avena-sterol	Other	Total phytosterols
Non-chocolate confectionery							
476 **Cereal crunchy bar**	0	6.9	6.9	27.2	4.0	0.9	45.8
479 **Liquorice**	0.5	0.9	0	2.4	0	0	3.8
Savoury snacks							
485 **Corn and starch snacks**	0	12.0	4.0	31.0	0	0	47.0
486 **Maize and rice flour snacks**	4.0	32.0	0	50	0	0	86.0
487 **Mixed cereal and potato snacks**	0	11.0	4.0	30	0	0	45.0
488 **Pork scratchings**	0	2.2	0	2.7	0	0	4.9
489 **Poppadums**, takeaway	18.7	75.3	11.4	116.6	4.7	6.4	233.1
492 **Potato and corn sticks**	4.0	20	2.0	39.0	2.0	0	67.0
494 **Potato rings**	0	4.0	0	11.0	0	0	15.0
495 **Prawn crackers**, takeaway	3.3	11.7	0.2	16.8	0.5	1.6	34.1
497 **Puffed potato products**	9.0	49.0	0	61.0	0	0	119.0
498 **Tortilla chips**	0	16.4	7.6	68.0	5.3	1.2	98.4
Beverages							
502 **Drinking chocolate**	0	1.3	2.9	8.6	0	0	12.8
503 **Instant drinks powder**, chocolate, low calorie	0	0	0	2.8	0	0	2.8
504 malted	0	0	0	3.7	0	0	3.7
Soups							
505 **Canned cream of chicken soup**, as served	1.0	6.0	0	8.0	0	0	15.0
506 **Canned cream of mushroom soup**, as served	1.0	6.0	0	8.0	0	0	15.0
507 **Canned cream of tomato soup**, as served	1.0	6.0	1.0	8.0	0	0	16.0
508 **Dried cream of tomato soup**	0	6.0	6.0	13.0	0	0	25.0
509 **Dried instant soup**	0	5.0	4.0	9.0	0	0	18.0
Sauces and dressings							
511 **Cook-in sauces**	1.5	7.1	0	9.7	0.7	0	19.0
512 **Dressing**, blue cheese	0.5	25.4	17.3	58.4	2.0	1.5	105.2
515 thousand island	11.4	64.2	9.0	99.0	7.2	1.8	192.6
516 tofu	18.4	235.8	132.3	546.5	21.5	11.1	965.5
517 **Horseradish sauce**	1.3	6.6	0.6	14.0	0.7	0.1	23.3
518 **Mayonnaise**, reduced calorie	0	9.3	7.6	32.5	0	3.3	52.7
520 **Salad cream**	0	12.3	8.9	71.8	0	2.9	95.8
521 reduced calorie	0	5.8	2.4	25.8	0	0	34.1

References to tables

van Poppel, G., van Erp-Baart, M., Leth, T, Gevers, E., van Amelsvoort, J., Lanzmann-Petithory, D., Kafatos, A., Aro, A. (1998) Trans fatty acids in foods in Europe: the TRANSFAIR study. *Journal of Food Composition and Analysis*, **11**: 112-136

Aro, A., van Amelsvoort, J., Becker, W., van Erp-Baart, M., Kafatos, A., Leth,T., van Poppel, G. (1998) Trans fatty acids in dietary fats and oils from 14 European countries: the TRANSFAIR study. *Journal of Food Composition and Analysis*, **11**: 137-149

Aro, A., Pizzoferrato, L., Reykdal, O., van Poppel, G. (1998) Trans fatty acids in dairy and meat products from 14 European countries: the TRANSFAIR study. *Journal of Food Composition and Analysis*, **11**: 150-160

van Erp-Baart, M., Couet, C. Cuadrado, C., Kafatos, A., Lanzmann, D., Stanley, J., van Poppel, G. (1998) Trans fatty acids in bakery products from 14 European countries: the TRANSFAIR study. *Journal of Food Composition and Analysis*, **11**: 161-169

Aro, A., Amaral, E., Kesteloot, H., Thamm, M., Rimestad, A., van Poppel, G. (1998) Trans fatty acids in french fries, soups and snacks from 14 European countries: the TRANSFAIR study. *Journal of Food Composition and Analysis*, **11**: 170-177

FOOD INDEX

Foods are indexed by their food publication number and **not** by their page number. The publication number is a number assigned to each food for ease of reference for the purpose of this supplement only.

The index includes two kinds of cross-reference. The first is the normal coverage of alternative names (e.g. Back bacon see **Bacon rashers, back**). The second is to common examples of components of generically described foods, including brand names, which although not part of the food name have in general been included in the product description (e.g. Anchor half fat butter see **Blended spread, 40% fat**).

The index also contains the old food code and a new food code where applicable (see Introduction). It is these food codes that will be used in nutrient databank applications.

	Publication number	New food code	Old food code
Ackee, canned, drained	420		13-145
All-Bran	19	11-364	11-126
Almonds	446	14-855	14-801
Anchor half fat butter	see **Blended spread, 40% fat**		
Animals	see **Shortcake, chocolate**		
Avocado	442	14-852	14-037
Bacon loin steaks, grilled	299		19-019
Bacon rashers, back, grilled	293		19-003
back, raw	292		19-001
middle, fried in corn oil	295		19-014
middle, grilled	296		19-015
middle, raw	294		19-013
streaky, grilled	298		19-018
streaky, raw	297		19-016
Bacon Streaks	see **Maize and rice flour snacks**		
Back bacon	see **Bacon rashers, back**		
Baked beans, canned in tomato sauce	421		13-043
Bakewell tart, individual	74	11-426	11-283
Ballisto	see **Chocolate bar with wafer/ biscuit and fruit**		
Balti, vegetable, takeaway	see **Vegetable balti, takeaway**		
Banana	443	14-853	14-045

	Publication number	New food code	Old food code
Barley, pearl, raw	1		11-002
Barnstormers	see **Oat based biscuits**		
Beanburgers	see **Vegetable grill/burger**		
Beans, baked, canned in tomato sauce	see **Baked beans**		
Beefburgers, chilled/frozen, fried in vegetable oil	307		19-029
raw	306		19-028
Beefburgers, economy, grilled	310		19-043
Beefburgers, reduced fat, chilled/frozen, fried in			
vegetable oil	309		19-036
raw	308		19-035
Beef, average, fat only, cooked	209	18-458	18-005
lean only, cooked	208		18-444
trimmed fat, raw	207	18-431	18-003
trimmed lean, raw	206	18-430	18-001
Beef bourguignonne, chilled/frozen, reheated	see **Beef in sauce with vegetables**		
Beef, braised steak	see **Beef in sauce with vegetables**		
Beef, braising steak, braised, lean and fat	see **Braising steak, beef**		
Beef, brisket, boiled, lean only	see **Brisket, beef**		
Beef, corned	see **Corned beef**		
Beef, dripping	see **Dripping, beef**		
Beef, fillet steak, cooked from steakhouse,			
lean only	see **Fillet steak, beef**		
Beef, flank, boneless, pot roasted, lean and fat	see **Flank, beef, boneless**		
Beef goulash	see **Beef in sauce with vegetables**		
Beef, grillsteaks, raw	see **Grillsteaks, beef**		
Beef hotpot with potatoes, chilled/frozen,			
reheated	see **Lamb/beef hotpot**		
Beef in sauce with vegetables, chilled/frozen,			
reheated	359		19-172
Beef, minced	see **Minced beef**		
Beef pancakes	see **Pancakes, beef**		
Beef pie	see also **Steak and kidney pie**		
Beef pie, chilled/frozen, baked	312	19-287	19-051
Beef, rump steak, fried in corn oil, lean only	see **Rump steak, beef**		
sirloin steak, grilled, medium rare, lean only	see **Sirloin steak, beef**		
Beef sausages, fried in corn oil	322		19-076
grilled	323		19-077
raw	321		19-075
Beef stew, meat only	358	19-297	
Beef stock cubes	see **Stock cubes, beef**		
Beef, stewing steak, stewed, lean and fat	see **Stewing steak, beef**		
Beef, stir-fried with green peppers in black			
bean sauce, takeaway	360	19-305	
Beef, topside, roasted, medium-rare, lean and fat	see **Topside, beef**		

	Publication number	New food code	Old food code
Belly joint/slices, pork, roasted, lean and fat	243		18-208
Best end neck cutlets, lamb, barbecued, lean only	228		18-103
Bhuna, prawn	see **Prawn bhuna**		
Biryani, vegetable, takeaway	see **Vegetable biryani, takeaway**		
Biscuits, Animals	see **Shortcake, chocolate**		
barnstormers	see **Oat based biscuits**		
Blue Riband	see **Wafers, filled, chocolate**		
bourbon	see **Sandwich biscuits, cream filled**		
Breakaway	see **Chocolate biscuits, full coated**		
cheese sandwich	see **Cheese sandwich biscuits**		
chocolate, cream filled, full coated	see **Chocolate biscuits, cream filled**		
chocolate, full coated	see **Chocolate biscuits, full coated**		
chocolate chip cookies	see **Chocolate chip cookies**		
chocolate fingers	see **Chocolate biscuits, full coated**		
chocolate wafers, filled, full coated	see **Wafers, filled, chocolate**		
Club	see **Chocolate biscuits, cream filled**		
Cornish wafers	see **Cornish wafers**		
cream crackers	see **Cream crackers**		
crunch, cream filled	see **Crunch biscuits**		
custard creams	see **Sandwich biscuits, cream filled**		
digestives	see **Digestives**		
fig rolls	see **Fig rolls**		
fruit	see **Fruit biscuits**		
fruit shortcakes	see **Fruit biscuits**		
gingernut	see **Gingernut biscuits**		
Hob Nob bars	see **Chocolate biscuits, cream filled**		
Hob Nob biscuits	see **Oat based biscuits**		
jaffa cakes	see **Jaffa cakes**		
jam rings	see **Sandwich biscuits, jam filled**		
Jammy Dodgers	see **Sandwich biscuits, jam filled**		
Jaspers	see **Fruit biscuits**		
Krackerwheat	see **Krackerwheat**		
Lincoln	see **Short sweet biscuits**		
Magic Numbers	see **Shortcake, chocolate**		
marie	see **Semi-sweet biscuits**		
oat based	see **Oat based biscuits**		
oatcakes	see **Oatcakes**		
osborne	see **Semi sweet biscuits**		
Penguin	see **Chocolate biscuits, cream filled**		
Rich Tea biscuits	see **Semi-sweet biscuits**		
sandwich, cream filled	see **Sandwich biscuits, cream filled**		
sandwich, jam filled	see **Sandwich biscuits, jam filled**		
semi-sweet	see **Semi-sweet biscuits**		

	Publication number	New food code	Old food code
short sweet biscuits	see **Short sweet biscuits**		
shortcake	see **Short sweet biscuits**		
shortcake, chocolate, half coated	see **Shortcake, chocolate**		
shorties	see **Shortcake, chocolate**		
Shrewsburys	see **Fruit biscuits**		
Signature	see **Shortcake, chocolate**		
Snapjacks	see **Oat based biscuits**		
Taxi	see **Wafers, filled, chocolate**		
Trio	see **Chocolate biscuits, cream filled**		
United	see **Chocolate biscuits, full coated**		
wafer	see **Cornish wafers**		
wafers, filled, chocolate, full coated	see **Wafers, filled, chocolate**		
water	see **Water biscuits**		
Bitza Pizza	see **Potato and tapioca snacks**		
Black Forest gateau	see **Gateau, frozen, chocolate based**		
Black pudding, raw	336		19-113
Blackcurrant seed oil	186	17-434	
Blended spread, 40% fat	174	17-455	17-016
Blended spread, 70-80% fat	173		17-015
Blue cheese dressing	see **Dressing, blue cheese**		
Blue Riband	see **Wafers, filled, chocolate**		
Bolognese, spaghetti, chilled, reheated	see **Spaghetti bolognese**		
Bombay mix	482		14-807
Borage oil	187	17-435	
Bounty bar	461		17-082
Bounty ice cream bar	see **Ice cream bar**		
Bourbon biscuits	see **Sandwich biscuits, cream filled**		
Bovril stock cubes	see **Stock cubes, beef**		
Braised steak, beef	see **Beef in sauce with vegetables**		
Braising steak, beef, braised, lean and fat	210		18-009
Bran Flakes	20	11-365	11-128
Brazil nuts	447	14-856	14-808
Bread, ciabatta	see **Bread, speciality, white**		
focaccia	see **Bread, speciality, white**		
pugliese	see **Bread, speciality, white**		
softgrain	14	11-398	
speciality, white	15	11-399	
white, average	12		11-099
wholemeal, average	13	11-363	11-113
Breaded chicken, stuffed with cheese and vegetables, chilled/frozen, baked	see **Chicken in crumbs**		
Breadsticks	483		17-123
Breakaway	see **Chocolate biscuits, full coated**		
Breakfast milk, pasteurised, average	89	12-885	

	Publication number	New food code	Old food code
Brie	128	12-861	12-131
Brisket, beef, boiled, lean only	211		18-014
Brown rice, raw	see **Rice, brown, raw**		
Burgers, bean	see **Beanburgers**		
Burgers, beef	see **Beefburgers**		
Butter	171		17-013
Butter, half fat	see **Blended spread, 40% fat**		
Butter, spreadable	172		17-014
Cake, cup, chocolate	see **Chocolate cup cake**		
fruit	see **Fruit cake**		
Madeira	see **Madeira cake**		
sponge, butter cream	see **Sponge cake**		
Cakes, fancy iced	see **Fancy iced cakes**		
Cakes, jaffa	see **Jaffa cakes**		
Canned cream of chicken soup, as served	505		17-252
Canned cream of mushroom soup, as served	506		17-270
Canned cream of tomato soup, as served	507		17-278
Cannelloni, chilled/frozen, reheated	361		19-184
Caramels, chocolate covered	see **Chocolate covered caramels**		
Casserole, beef, meat only	see **Beef stew**		
Cereal chewy bar	475	17-428	17-102
Cereal crunchy bar	476		17-103
Cheddar, average	129	12-862	12-134
Cheesecake, fruit, individual	151	12-911	
Cheese and onion crispbakes	see **Vegetable and cheese grill/burger**		
Cheese, Brie	see **Brie**		
Cheddar	see **Cheddar**		
Cottage	see **Cottage cheese**		
Cream	see **Cream cheese**		
Edam	see **Edam**		
slices, processed	see **Processed cheese slices**		
stilton, blue	see **Stilton**		
Cheese grills	see **Vegetable and cheese grill/burger**		
Cheese nachos, takeaway	79	11-428	
Cheese sandwich biscuits	31	11-404	
Chewits	see **Chew sweets**		
Chew sweets	477		17-104
Chewy cereal bar	see **Cereal chewy bar**		
Chick peas, hummus	422	13-863	13-088
Chicken and ham pie	see **Chicken pie**		
Chicken and mushroom pie	see **Chicken pie**		
Chicken and vegetable pie	see **Chicken pie**		
Chicken in crumbs, stuffed with cheese and vegetables, chilled/frozen, baked	337		19-116

	Publication number	New food code	Old food code
Chicken chop suey, takeaway	362	19-298	
Chicken creole	see **Chicken in sauce with vegetables**		
Chicken eggs	160	12-879	12-801
Chicken fried rice, takeaway	363	19-299	
Chicken in sauce with vegetables, chilled/frozen, reheated	364		19-193
Chicken in white sauce, canned	365		19-194
Chicken kiev, frozen, baked	338		19-123
Chicken korma, takeaway	366	19-300	
Chicken, breast, grilled, meat only	263		18-323
Chicken, dark meat, raw	257		18-289
roasted	261		18-330
Chicken, fresh, corn-fed, whole, raw, dark meat	267	18-453	18-299
light meat	268	18-452	18-300
Chicken, fresh, portions, casseroled with skin,			
dark meat	264	18-450	
light meat	265	18-449	
Chicken, light meat, raw	258		18-290
roasted	262		18-329
Chicken pancakes	see **Pancakes, chicken**		
Chicken pie, individual, chilled/frozen, baked	313	19-288	19-055
Chicken portions, fried, meat only, takeaway	266	18-451	
Chicken roll, cooked	339		19-125
Chicken, skin, cooked	260	18-448	
raw	259		18-292
Chicken soup, cream of, canned	see **Canned cream of chicken soup**		
Chicken tandoori, chilled, reheated	340		19-127
Chicken tikka masala, takeaway	367	19-301	
Chicken tikka, chilled, reheated	341	19-296	
Chicken with cashew nuts, takeaway	368	19-302	
Chicken, tomato and mushroom casserole	see **Chicken in sauce with vegetables**		
Chinese style crackers	see **Puffed potato products**		
Chips, microchips, microwaved	see **Microchips**		
Chips, potato	see **Potato chips**		
Chipsticks	see **Potato and corn sticks**		
Chocolate bar with wafer/biscuit and fruit	463		17-084
Chocolate biscuits, cream filled, full coated	33	11-405	
Chocolate biscuits, full coated	32	11-373	11-166
Chocolate buttons	see **Chocolate, milk**		
Chocolate chip cookies	34	11-406	
Chocolate covered caramels	462		17-083
Chocolate covered marshmallow teacake	56	11-419	
Chocolate cup cake	55	11-418	
Chocolate dairy dessert, individual	152	12-912	

	Publication number	New food code	Old food code
Chocolate digestives	see **Digestives, chocolate**		
Chocolate fingers	see **Chocolate biscuits, full coated**		
Chocolate flavoured milk, pasteurised	96		12-891
Chocolate gateau	see **Gateau, frozen, chocolate based**		
Chocolate instant drinks powder	see **Instant drinks powder**		
Chocolate mousse, individual	see **Mousse, chocolate**		
Chocolate mousse, reduced fat	see **Mousse, chocolate**		
Chocolate nut spread	458	17-422	17-070
Chocolate shortcake	see **Shortcake, chocolate**		
Chocolate swiss roll	see **Swiss roll, chocolate**		
Chocolate trifle, individual	see **Trifle, chocolate**		
Chocolate wafers, filled, full coated	see **Wafers, filled, chocolate**		
Chocolate, cooking	see **Cooking chocolate**		
Chocolate, couverture	see **Couverture chocolate**		
Chocolate, fancy and filled	466		17-088
Chocolate, milk	467		17-089
Chocolate, plain	468		17-090
Chop suey, chicken, takeaway	see **Chicken chop suey**		
Chops, loin, lamb, grilled, lean only	see **Loin chops, lamb**		
Chops, loin, pork, microwaved, lean and fat	see **Loin chops, pork, microwaved**		
Chops, loin, pork, grilled, lean and fat	see **Loin chops, pork, grilled**		
Chump chops/steaks, pork, fried in corn oil, lean and fat	244		18-215
Chump steaks/chops, lamb, fried in corn oil, lean only	229		18-117
Ciabatta bread	see **Bread, speciality, white**		
Clover	see **Fat spread, 70-80% fat, not polyunsaturated**		
Club	see **Chocolate biscuits, cream filled**		
Coconut oil	188		17-031
Coconut, fresh	448	14-868	14-816
Cod liver oil	189		17-032
Cod, raw	381		16-012
Cod, roe, hard, raw	see **Roe, cod, hard, raw**		
Coffee compliment powder	92	12-836	12-026
Coffee whitener liquid, with glucose syrup and vegetable fat	95		12-889
Coffee whitener liquid, with skimmed milk and non milk fat	94		12-888
Coffee whitener powder, low fat	93		12-887
Compliment, coffee powder	see **Coffee compliment powder**		
Compound cooking fat	165		17-471
Cook-in sauces	511	17-479	17-295
Cookeen	see **Compound cooking fat**		
Cookies, chocolate chip	see **Chocolate chip cookies**		

	Publication number	New food code	Old food code
Cooking chocolate	464		17-086
Cooking fat, compound	see **Compound cooking fat**		
Corn and starch snacks	485	17-429	17-126
Corn oil	190	17-419	17-033
Corn snacks	484		17-125
Corned beef	342		19-128
Cornflakes	21	11-366	11-130
Cornflakes, Crunchy Nut	see **Crunchy Nut Corn Flakes**		
Cornish pastie, cooked	314		19-056
Cornish wafers	35	11-407	
Cottage cheese	130	12-900	
Cottage/Shepherd's pie, chilled/frozen, reheated	369		19-216
Cottonseed oil	191		17-034
Couverture chocolate	465	17-453	
Cow & Gate Infasoy, reconstituted	124	12-857	12-063
Cow & Gate Plus, reconstituted	120	12-853	12-053
Cow & Gate Premium, reconstituted	117	12-850	12-047
Crab, boiled	397		16-232
Crackers, cream	see **Cream crackers**		
prawn	see **Prawn crackers**		
Cream cheese	131	12-863	12-150
Cream crackers	36	11-374	11-167
Cream, dairy, UHT, canned spray	see **Dairy cream, UHT**		
double, fresh, pasteurised	see **Double cream**		
Cream liqueur, whisky based	501	17-449	17-242
Cream of chicken soup	see **Canned cream of chicken soup**		
Cream of mushroom soup	see **Canned cream of mushroom soup**		
Cream of tomato soup	see **Canned cream of tomato soup**		
Cream, single, fresh, pasteurised	see **Single cream**		
Creme caramel	153	12-875	12-220
Creme eggs	469		17-092
Creme fraiche	107	12-896	
Creole, chicken	see **Chicken in sauce with vegetables**		
Crispbakes	see **Vegetable grill/burger**		
Crispbakes, cheese and onion	see **Vegetable and cheese grill/burger**		
Crisps, potato	see **Potato crisps**		
Crispy Caramel	see **Chocolate bar with wafer/biscuit and fruit**		
Croissants, plain, retail	16	11-400	
Croissants, savoury, retail	17	11-401	
Croissants, sweet, retail	18	11-402	
Crunch biscuits, cream filled	37	11-408	
Crunchy cereal bar	see **Cereal crunchy bar**		
Crunchy Nut Corn Flakes	22	11-367	11-131
Crunchy Oat Cereal	23	11-403	

	Publication number	New food code	Old food code
Crunchy sticks	see **Potato and corn sticks**		
Cumberland sausages	see **Premium sausages**		
Cup cake, chocolate	see **Chocolate cup cake**		
Custard creams	see **Sandwich biscuits, cream filled**		
Custard, ready to eat	154	12-876	12-225
Custard tarts, individual	68	11-390	11-239
Dairy cream, UHT, canned spray	108	12-844	12-124
Dairy dessert, chocolate, individual	see **Chocolate dairy dessert**		
Dairy Milk	see **Chocolate, milk**		
Danish pastries	69	11-391	11-240
Delight Diet	see **Fat spread, 20-25% fat, not polyunsaturated**		
Delight, low fat spread	see **Fat spread, 40% fat, not polyunsaturated**		
Diced pork, casseroled, lean only	246		18-219
Diced pork, raw, lean only	245	18-447	18-217
Diet yogurt/virtually fat free, fruit, twin pot	see **Virtually fat free/diet yogurt**		
Digestives, chocolate, half coated	38	11-375	11-169
Digestives, plain	39	11-376	11-170
Doner kebabs, cooked, lean only	343		19-129
Double cream, fresh, pasteurised	106	12-843	12-116
Doughnuts, ring	70	11-392	11-243
Dressing, blue cheese	512	17-480	17-300
Dressing, French	513	17-481	17-302
Dressing, reduced calorie	514	17-450	
Dressing, thousand island	515	17-482	17-306
Dressing, tofu	516	17-451	
Dried cream of tomato soup	508		17-281
Dried instant soup	509	17-431	17-259
Drinking chocolate	502	12-883	12-093
Dripping, beef	166	17-472	17-006
Duck eggs	161	12-880	12-813
Duckling, raw, meat only	275	18-461	18-369
fat and skin	276	18-462	18-371
Duckling, roasted, meat only	277	18-463	18-372
meat, fat and skin	278	18-464	18-374
Edam	132	12-864	12-154
Eel, jellied	384		16-174
Egg fried rice, takeaway	163	12-916	
Eggs, chicken	see **Chicken eggs**		
Eggs, duck	see **Duck eggs**		
Eggs, quail	see **Quail eggs**		
Eggs, scotch	see **Scotch eggs**		
Elmlea, double	111	12-847	12-128
Elmlea, single	109	12-845	12-126
Elmlea, whipping	110	12-846	12-127

	Publication number	New food code	Old food code
Enchiladas, vegetable, takeaway	424	15-352	
Evaporated milk, light	91	12-890	
Evening primrose oil	192	17-420	17-035
Faggots in gravy	344		19-131
Fancy iced cakes	57	11-383	11-199
Farley's First Milk, reconstituted	118	12-851	12-048
Farley's Second Milk, reconstituted	121	12-854	12-055
Farley's Soya Formula, reconstituted	125	12-858	12-061
Fat spread, 20-25% fat, not polyunsaturated	185	17-476	17-028
Fat spread, 35-40% fat, polyunsaturated	184	17-475	17-027
Fat spread, 40% fat, not polyunsaturated	183	17-474	17-026
Fat spread, 60% fat, with olive oil	182		17-025
Fat spread, 70% fat, monounsaturated	180	17-460	
Fat spread, 70% fat, polyunsaturated	181	17-470	17-023
Fat spread, 70-80% fat, not polyunsaturated	179	17-433	17-022
Fig rolls	40	11-409	
Fillet steak, beef, cooked from steakhouse, lean only	212		18-022
Fillet strips, pork, stir-fried, lean only	247	18-228	
Fish cakes, frozen, raw	404	16-322	16-280
Fish fingers, cod, frozen, raw	405	16-312	16-287
Flank, beef, boneless, pot roasted, lean and fat	213		18-027
Flavoured milk, pasteurised	97	12-837	12-034
Flora	see **Fat spread, 70% fat, polyunsaturated**		
Flora Light	see **Fat spread, 35-40% fat, polyunsaturated**		
Flora, white	see **Compound cooking fat**		
Flour, rye, whole	see **Rye flour**		
Flour, soya	see **Soya flour**		
Flour, wheat, brown	see **Wheat flour, brown**		
white, household, plain	see **Wheat flour, white, household, plain**		
wholemeal	see **Wheat flour, wholemeal**		
Focaccia bread	see **Bread, speciality, white**		
Fondant fancies	see **Fancy iced cakes**		
Fool, fruit	see **Fruit fool**		
Frankfurter	332		19-100
Frazzles	see **Maize and rice flour snacks**		
French dressing	see **Dressing, French**		
French fancies	see **Fancy iced cakes**		
Fritters, potato	see **Potato fritters**		
Fromage frais, fruit, children's	145	12-908	
Fromage frais, plain	144	12-871	12-158
Fruit biscuits	41	11-410	
Fruit cake, plain	58	11-384	11-200
Fruit cake, rich	59	11-385	11-201

	Publication number	New food code	Old food code
Fruit cheesecake, individual	see **Cheesecake, fruit**		
Fruit fool, individual	155	12-913	
Fruit fromage frais	see **Fromage frais, fruit**		
Fruit gateau	see **Gateau, frozen, fruit**		
Fruit 'n Fibre	see **Optima/Fruit 'n Fibre**		
Fruit pies, individual, double crust	75	11-394	11-310
Fruit shortcakes	see **Fruit biscuits**		
Fruit torte, frozen/chilled	see **Torte, frozen/chilled**		
Fruit trifle	see **Trifle, fruit**		
Fruitella	see **Chew sweets**		
Fudge	478	17-444	
Fusilli, plain, fresh, cooked	see **Pasta, plain**		
Galaxy	see **Chocolate, milk**		
Gammon	see **Ham, gammon**		
Gateau, frozen, chocolate based	60	11-421	
Gateau, frozen, fruit	61	11-420	
Ghee, vegetable	167		17-009
Ginger, ground	440	13-865	13-832
Gingernut biscuits	42	11-377	11-172
Goats' milk	98	12-838	12-037
Gold Light	see **Fat spread, 40% fat, not polyunsaturated**		
Gold Lowest	see **Fat spread, 20-25% fat, not polyunsaturated**		
Goulash, beef	see **Beef in sauce with vegetables**		
Grapeseed oil	193	17-436	17-036
Greek style yogurt, plain	146	12-872	12-194
Grill/burger, protein substitute, unbreaded, baked/grilled	see **Protein substitute grill/burger**		
Grillsteaks, beef, raw	311	19-307	19-044
Haddock, raw	382		16-044
Ham, regular	302		19-023
Ham and mushroom tagliatelle	see **Tagliatelle with ham and mushroom**		
Ham and pork, chopped, canned	345		19-133
Ham, gammon joint, boiled	300		19-020
Ham, gammon rashers, grilled	301		19-022
Ham, premium	304		19-026
Hazelnut oil	194	17-437	17-037
Hazelnut yogurt, low fat	see **Low fat yogurt, hazelnut**		
Hazelnuts	449	14-869	14-821
Heart, lamb, raw	284	18-438	18-396
Heart, pig, raw	285	18-439	18-400
Herring, raw	385	16-309	16-175
Herring, roe, soft, raw	see **Roe, herring**		
Hob Nob bars	see **Chocolate biscuits, cream filled**		
Hob Nob biscuits	see **Oat based biscuits**		

	Publication number	New food code	Old food code
Horseradish sauce	517		17-314
Hot cross buns	see **Teacakes**		
Hotpot, lamb/beef with potatoes, chilled/frozen, reheated	see **Lamb/beef hotpot**		
Hula Hoops	see **Potato rings**		
Human milk, mature	99	12-839	12-040
Hummus, chick peas	see **Chick peas, hummus**		
I Can't Believe It's Not Butter!	see **Fat spread, 70% fat, monounsaturated**		
Ice cream bar, chocolate coated	150	12-910	
Ice cream, dairy	148	12-874	12-204
Ice cream, reduced calorie	149	12-909	
Infasoy, Cow & Gate	see **Cow & Gate Infasoy**		
Instant drinks powder, chocolate, low calorie	503	12-917	
Instant drinks powder, malted	504	12-884	
Instant soup, dried	see **Dried instant soup**		
Irish stew, canned	371		19-234
Jaffa cakes	43	11-378	11-177
Jam rings	see **Sandwich biscuits, jam filled**		
Jam tarts, retail	71	11-393	11-255
Jammy Dodgers	see **Sandwich biscuits, jam filled**		
Jaspers	see **Fruit biscuits**		
Kebabs, doner, cooked, lean only	see **Doner kebabs**		
Kebabs, shish, cooked, lean only	see **Shish kebabs**		
Kidney, lamb, grilled	286	18-456	
Kidney, pig, raw	287	18-440	18-406
Kiev, chicken, frozen, baked	see **Chicken kiev**		
Kiev, vegetable, baked	see **Vegetable kiev**		
Kippers, raw	386		16-187
Kit Kat	470	17-424	17-093
Korma, chicken, takeway	see **Chicken korma**		
Krackerwheat	44	11-411	
Krona	see **Fat spread, 70-80% fat, not polyunsaturated**		
Lamb roasts, frozen, cooked	346		19-134
Lamb rogan josh, takeaway	372	19-306	
Lamb samosa, cooked	316	19-289	19-059
Lamb, average, fat only, cooked	227	18-434	18-100
lean only, cooked	226	18-445	
trimmed fat, raw	225	18-433	18-098
trimmed lean, raw	224	18-432	18-096
Lamb, best end neck cutlets, barbecued, lean only	see **Best end neck cutlets, lamb**		
Lamb, chump steaks/chops, fried in corn oil, lean only	see **Chump steaks/chops, lamb**		
Lamb, heart, raw	see **Heart, lamb**		

	Publication number	New food code	Old food code
Lamb, kidney, grilled	see **Kidney, lamb**		
Lamb, leg, average, raw, lean and fat	see **Leg, average, lamb**		
Lamb, liver, raw	see **Liver, lamb**		
Lamb, loin chops, grilled, lean only	see **Loin chops, lamb**		
Lamb, minced, stewed, lean and fat	see **Minced lamb**		
Lamb, neck fillet, strips, stir-fried in corn oil, lean only	see **Neck fillet, strips, lamb**		
Lamb, rack of, roasted, lean and fat	see **Rack of lamb**		
Lamb, shoulder, half bladeside, pot-roasted, lean and fat	see **Shoulder, lamb, half bladeside**		
Lamb, stewing, stewed, lean and fat	see **Stewing lamb**		
Lamb, leg, whole, roasted medium, lean and fat	see **Leg, whole, lamb**		
Lamb, shoulder, whole, roasted, lean and fat	see **Shoulder, whole, lamb**		
Lamb/beef hotpot with potatoes, chilled/frozen, reheated	370		19-231
Lard	168		17-010
Lasagne, chilled/frozen, reheated	373		19-238
Lasagne, plain, fresh, cooked	see **Pasta, plain**		
Lassi, sweetened	100	12-892	
Leg, average, lamb, raw, lean and fat	230		18-123
Leg joint, pork, roasted medium, lean and fat	248		18-241
Leg, whole, lamb, roasted medium, lean and fat	231		18-136
Lemon curd	459	17-423	17-076
Lincoln biscuits	see **Short sweet biscuits**		
Lincolnshire sausages	see **Premium sausages**		
Linguine, plain, fresh, cooked	see **Pasta, plain**		
Lion bar	see **Chocolate bar with wafer/biscuit and fruit**		
Liqueur, cream, whisky based	see **Cream liqueur**		
Liquorice	479	17-445	
Liquorice allsorts	480	17-446	17-112
Liver pâté	see **Pâté, liver**		
Liver, chicken, raw	288	18-441	18-411
Liver, lamb, raw	289	18-442	18-413
Liver, pig, raw	290	18-443	18-417
Loin chops, lamb, grilled, lean only	232		18-141
Loin chops, pork, microwaved, lean and fat	250		18-254
Loin chops, pork, grilled, lean and fat	249		18-252
Loin joint, pork, roasted, lean and fat	251		18-263
Loin steaks, pork, fried in corn oil, lean only	252		18-265
Low fat yogurt, hazelnut	140	12-904	
Low fat yogurt, plain	139	12-870	12-188
Low fat yogurt, toffee	141	12-905	
Luncheon meat, canned	347		19-135
Mackerel, raw	387		16-191

	Publication number	New food code	Old food code
Madeira cake	62	11-386	11-209
Madras, prawn	see **Prawn madras**		
Magic Numbers, biscuits	see **Shortcake, chocolate**		
Maize and rice flour snacks	486		17-127
Malted instant drinks powder	see **Instant drinks powder, malted**		
Margarine, catering	175	17-442	
Margarine, hard, vegetable fats only	176	17-456	17-019
Margarine, soft, not polyunsaturated	177	17-458	17-020
Margarine, soya	178	17-432	
Marie biscuits	see **Semi-sweet biscuits**		
Mars bar	471	17-425	17-094
Mars ice cream bar	see **Ice cream bar**		
Marshmallow teacake, chocolate covered	see **Chocolate covered marshmallow teacake**		
Marzipan	460	14-861	14-825
Mayonnaise	518		17-316
Mayonnaise, reduced calorie	519	17-483	17-318
Meat loaf, chilled/frozen, reheated	348		19-138
Meat pâté, low fat	see **Pâté, meat**		
Meat, luncheon, canned	see **Luncheon meat**		
Microchips, microwaved	417	13-869	13-028
Milk chocolate	see **Chocolate, milk**		
Milk, breakfast, pasteurised, average	see **Breakfast milk**		
Milk, chocolate flavoured, pasteurised	see **Chocolate flavoured milk**		
Milk, evaporated, light	see **Evaporated milk, light**		
Milk, flavoured, pasteurised	see **Flavoured milk, pasteurised**		
Milk, goats'	see **Goats' milk**		
Milk, human, mature	see **Human milk**		
Milk, organic, semi-skimmed, pasteurised	see **Organic semi-skimmed milk**		
Milk, semi-skimmed, pasteurised, average	see **Semi-skimmed milk**		
Milk, sheep's	see **Sheep's milk**		
Milk, skimmed, pasteurised, average	see **Skimmed milk**		
Milk, whole, pasteurised, average	see **Whole milk, pasteurised, average**		
sterilised	see **Whole milk, sterilised**		
UHT	see **Whole milk, UHT**		
Milkshake, thick, takeaway	101		12-893
Milky way	472	17-426	17-095
Milupa Aptamil, reconstituted	116	12-849	12-045
Milupa Milumil, reconstituted	122	12-855	12-057
Mince in gravy, canned	349		19-140
Minced beef, extra lean, raw	216		18-040
Minced beef, raw	214		18-036
Minced beef, stewed	215		18-038
Minced lamb, stewed, lean and fat	233		18-159
Minced pork, dry fried and stewed, lean and fat	256		18-268

	Publication number	New food code	Old food code
Minced pork, raw, lean and fat	255		18-267
Mixed cereal and potato snacks	487	17-430	17-128
Monster Munch	see **Corn snacks**		
Moussaka, chilled/frozen/longlife, reheated	374		19-248
Mousse, chocolate, individual	156	12-877	12-244/24
Mousse, chocolate, reduced fat, individual	157	12-914	
Muesli	25	11-369	11-137
Mushroom soup, cream of, canned	see **Canned cream of mushroom soup**		
Mussels, boiled	398		16-256
Mustard powder	441	13-866	13-838
Mycoprotein, Quorn, raw	see **Quorn mycoprotein**		
Nachos, cheese, takeaway	see **Cheese nachos**		
Neck cutlets, lamb, barbecued, lean only	see **Best end neck cutlets, lamb**		
Neck fillet, strips, lamb, stir-fried in corn oil, lean only	234		18-164
Nik Naks	see **Corn snacks**		
Non-dairy cream, UHT, canned spray	112	12-897	
Nuts and raisins, yogurt coated	see **Yogurt coated nuts and raisins**		
Nuts, almonds	see **Almonds**		
Brazil	see **Brazil nuts**		
coconut	see **Coconut**		
hazel	see **Hazelnuts**		
peanuts	see **Peanuts**		
Shanghai	see **Shanghai nuts**		
walnuts	see **Walnuts**		
Oat based biscuits	46	11-412	
chocolate, half coated	47	11-413	
Oat cereal, crunchy	see **Crunchy oat cereal**		
Oatbakes	see **Oat based biscuits**		
Oatcakes	45	11-379	11-181
Oatmeal, quick cook, raw	2		11-018
Oil, blackcurrant seed	see **Blackcurrant, seed oil**		
Oil, borage	see **Borage oil**		
Oil, coconut	see **Coconut oil**		
Oil, cod liver	see **Cod liver oil**		
Oil, corn	see **Corn oil**		
Oil, cottonseed	see **Cottonseed oil**		
Oil, evening primrose	see **Evening primrose oil**		
Oil, grapeseed	see **Grapeseed oil**		
Oil, hazelnut	see **Hazelnut oil**		
Oil, olive	see **Olive oil**		
Oil, palm	see **Palm oil**		
Oil, peanut	see **Peanut oil**		
Oil, rapeseed	see **Rapeseed oil**		

	Publication number	New food code	Old food code
Oil, safflower	see **Safflower oil**		
Oil, sesame	see **Sesame oil**		
Oil, soya	see **Soya oil**		
Oil, sunflower	see **Sunflower oil**		
Oil, vegetable, blended	see **Vegetable oil, blended**		
Oil, walnut	see **Walnut oil**		
Oil, wheatgerm	see **Wheatgerm oil**		
Olive oil	195		17-038
Olives	445	14-854	14-173
Olivio	see **Fat spread, 60% fat, with olive oil**		
Optima/Fruit 'n Fibre	24	11-368	11-134
Organic semi-skimmed milk, pasteurised	90	12-886	
Osborne biscuits	see **Semi- sweet biscuits**		
Oxo stock cubes	see **Stock cubes**		
Oysters, raw	399	16-311	16-260
Palm oil	196		17-039
Pancakes, beef, frozen, shallow-fried in vegetable oil	375		19-249
Pancakes, chicken, frozen, shallow-fried in vegetable oil	376		19-250
Pasta, plain, fresh, cooked	11	11-397	
Pastie, Cornish, cooked	see **Cornish pastie**		
Pastries, Danish	see **Danish pastries**		
Pastry, puff, frozen	see **Puff pastry**		
shortcrust, frozen	see **Shortcrust pastry**		
Pâté, liver	350		19-143
meat, low fat	351		19-145
vegetable	see **Vegetable pâté**		
Peanut butter, smooth	451	14-865	
Peanut oil	197		17-040
Peanuts, plain	450	14-857	14-832
Pearl barley, raw	see **Barley, pearl**		
Peas, raw	423	13-864	13-127
Penguin bars	see **Chocolate biscuits, cream filled**		
Pepperami	333		19-108
Pheasant casseroled, meat only	280	18-454	
Pheasant, roasted, meat only	279	18-465	18-383
Picnic	see **Chocolate bar with wafer/biscuit and fruit**		
Pie, beef, chilled/frozen, baked	see **Beef pie, chilled/frozen, baked**		
chicken, individual, chilled/frozen, baked	see **Chicken pie, individual, chilled/frozen, baked**		
Pie, pork, individual	see **Pork pie**		
Pie, steak and kidney, individual, cooked	see **Steak and kidney pie**		
Pies, fruit, individual, double crust	see **Fruit pies**		
Pig, heart, raw	see **Heart, pig**		

	Publication number	New food code	Old food code
Pig, liver, raw	see **Liver, pig**		
Pilchards, canned in tomato sauce	388		16-201
Pizza, deep pan, cheese and tomato, takeway	82	11-431	
meat topped, takeaway	83	11-432	
Pizza, thin base, cheese and tomato, takeaway	80	11-429	
fish topped, takeaway	81	11-430	
Plaice, raw	383		16-102
Plain chocolate	see **Chocolate, plain**		
Poppadums, takeaway	489	17-447	
Poppy seeds	452	14-866	
Pork and beef sausages, economy, raw	328		19-090
Pork haslet, cooked	352		19-146
Pork pie, individual	315		19-063
Pork roasts, frozen, cooked	353		19-147
Pork sausages, fried in corn oil	325		19-079
raw	324		19-078
reduced fat, fried in corn oil	327		19-085
reduced fat, raw	326		19-084
Pork scratchings	488	17-477	17-132
Pork shoulder, cured, slices	305		19-027
Pork steaks, grilled, lean only	254		18-285
Pork, average, fat only, cooked	242	18-437	18-203
lean only, cooked	241	18-446	
trimmed fat,raw	240	18-436	18-203
trimmed lean, raw	239	18-435	18-201
Pork, belly joint/slices, roasted, lean and fat	see **Belly joint/slices, pork**		
Pork, chump chops/steaks, fried in corn oil, lean and fat	see **Chump chops, steaks, pork**		
Pork, diced	see **Diced pork**		
Pork, fillet strips, stir-fried, lean only	see **Fillet strips, pork**		
Pork, leg joint, roasted, medium, lean and fat	see **Leg joint, pork**		
Pork, minced	see **Minced pork**		
Pork, loin chops, microwaved, lean and fat	see **Loin chops, pork, microwaved**		
grilled, lean and fat	see **Loin chops, pork, grilled**		
Pork, loin joint, roasted, lean and fat	see **Loin joint, pork**		
Pork, loin steaks, fried in corn oil, lean only	see **Loin steaks, pork**		
Pork, spare-ribs, 'barbecue style', chilled/frozen, reheated	377		19-262
Pork, spare ribs, sliced, grilled, lean and fat	see **Spare ribs, pork, sliced**		
Pork, sweet and sour, battered, takeaway	see **Sweet and sour pork**		
Potato and corn sticks	492		17-140
Potato and tapioca snacks	493		17-141
Potato chips, fast food chain	415	13-867	
Potato chips, fish and chip shop	416	13-868	13-022

	Publication number	New food code	Old food code
Potato crisps, plain	490		17-133
Potato crisps, reduced fat	491		17-136
Potato fritters, battered, cooked	418	13-870	
Potato rings	494		17-142
Potato waffles, baked	419	13-862	50-691
Potato, raw	414		13-009
Prawn bhuna, takeaway	406	16-317	
Prawn crackers, takeaway	495	17-448	
Prawn madras, takeaway	407	16-318	
Prawn toasts, sesame, takeaway	see **Sesame prawn toasts**		
Prawns, frozen, raw	400		16-241
Prawns, Szechuan, with vegetables, takeaway	see **Szechuan prawns**		
Premium sausages, fried in vegetable oil	330		19-094
raw	329		19-093
Processed cheese slices	133	12-866	12-172
reduced fat	134	12-902	
Prosobee, reconstituted	126	12-859	12-065
Protein substitute grill/burger, unbreaded, baked/grilled	425	15-353	
Pudding, black, raw	see **Black pudding**		
rice, canned	see **Rice pudding, canned**		
sponge, canned	see **Sponge pudding, canned**		
steak and kidney, canned	see **Steak and kidney pudding, canned**		
Puff pastry, frozen	66	11-388	11-224
Puffed potato products	497		17-147
Puffed Wheat	26		11-144
Pugliese bread	see **Bread, speciality, white**		
Punjabi puri	496		17-146
Pumpkin seeds	453		14-842
Quail eggs	162	12-881	12-815
Quavers	see **Puffed potato products**		
Quorn mycoprotein, raw	426	17-421	17-336
Rabbit, stewed, meat and fat	281	18-466	18-388
Rack of lamb, roasted, lean and fat	235		18-169
Rainbow trout	see **Trout, rainbow**		
Rapeseed oil	198		17-041
Ready Brek	27	11-370	11-145
Reduced fat yogurt, frozen	142	12-906	
Rice desserts, with fruit, individual, chilled	76	11-427	
Rice pudding, canned	77	11-395	11-324
Rice, brown, raw	9		11-035
Rice, white, easy cook, raw	10		11-042
Rich Tea biscuits	see **Semi-sweet biscuits**		
Ringos	see **Mixed cereal and potato snacks**		

	Publication number	New food code	Old food code
Rissoles, savoury, cooked	354		19-149
Roe, cod, hard, raw	408		16-299
Roe, herring, soft, raw	409		16-302
Rogan josh, lamb, takeaway	see **Lamb rogan josh**		
Roll, chicken, cooked	see **Chicken roll**		
Rump steak, beef, fried in corn oil, lean only	217		18-047
Rye flour, whole	6		11-022
Safflower oil	199	17-438	17-042
Salad cream	520	17-484	17-326
Salad cream, reduced calorie	521	17-452	17-327
Salami	334		19-110
Salmon, canned in brine, drained	389	16-310	16-210
Samosa, lamb, cooked	see **Lamb samosa**		
vegetable	see **Vegetable samosa**		
Sandwich biscuits, cheese	see **Cheese sandwich biscuits**		
cream filled	48	11-380	11-182
jam filled	49	11-414	
Sandwich spread	522	17-454	17-328
Sardines, canned in oil, drained	390	16-314	16-216
canned in tomato sauce	391		16-217
Sauce, horseradish	see **Horseradish sauce**		
Sauces, cook-in	see **Cook-in sauces**		
Sausages, beef	see **Beef sausages**		
pork	see **Pork sausages**		
pork and beef, economy, raw	see **Pork and beef sausages**		
premium	see **Premium sausages**		
turkey	see **Turkey sausages**		
vegetarian, baked/grilled	see **Vegetarian sausages**		
Sausage roll, flaky pastry, cooked	317	19-293	
Saveloy, unbattered, takeaway	335	19-292	19-111
Scampi, breaded, cooked	410	16-319	
Scones, retail	72	11-424	
Scotch eggs, retail	164	12-882	12-824
Seeds, poppy	see **Poppy seeds**		
pumpkin	see **Pumpkin seeds**		
sesame	see **Sesame seeds**		
sunflower	see **Sunflower seeds**		
Semi-sweet biscuits	50	11-381	11-183
Semi-skimmed milk, organic, pasteurised	see **Organic semi-skimmed milk**		
Semi-skimmed milk, pasteurised, average	85	12-832	12-008
Sesame oil	200		17-043
Sesame prawn toasts, takeaway	412	16-321	
Sesame seeds	454	14-858	14-844
Shanghai nuts	455	14-867	

	Publication number	New food code	Old food code
Sheep's milk	102	12-840	12-041
Shepherd's pie, chilled/frozen, reheated	see **Cottage/Shepherd's pie**		
Shish kebabs, cooked, lean only	355		19-150
Short sweet biscuits	52	11-382	11-184
Shortcake	see **Short sweet biscuits**		
Shortcake, chocolate, half coated	51	11-415	
Shortcrust pastry, frozen	67	11-389	11-227
Shorties	see **Shortcake, chocolate**		
Shoulder, lamb, half bladeside, pot-roasted, lean and fat	236		18-174
Shoulder, lamb, whole, roasted, lean and fat	237		18-180
Shredded Wheat	28	11-371	11-148
Shrewsburys	see **Fruit biscuits**		
Signature biscuits	see **Shortcake, chocolate**		
Simply Double dessert topping	113	12-898	
Single cream, fresh, pasteurised	105	12-842	12-113
Sirloin steak, beef, grilled medium rare, lean only	219		18-070
Skimmed milk, pasteurised, average	84	12-831	12-002
Skips	see **Corn and starch snacks**		
Sliverside, pot roasted, lean and fat only	218		18-057
SMA Gold, reconstituted	119	12-852	12-051
SMA White, reconstituted	123	12-856	12-059
SMA Wysoy, reconstituted	127	12-860	12-067
Snapjacks	see **Oat based biscuits**		
Snaps	see **Puffed potato products**		
Snickers ice cream bar	see **Ice cream bar**		
Softgrain bread	see **Bread, softgrain**		
Soup, cream of chicken, canned	see **Canned cream of chicken soup**		
Soup, cream of mushroom, canned	see **Canned cream of mushroom soup**		
Soup, cream of tomato, canned	see **Canned cream of tomato soup**		
Soya dessert topping	114	12-899	
Soya flour, full fat	7	11-361	11-025
Soya flour, reduced fat	8	11-362	11-026
Soya, non-dairy alternative to milk	103	12-841	12-042
Soya oil	201		17-044
Soya, alternative to yogurt, fruit	147	12-873	12-196
Spaghetti bolognese, chilled, reheated	378		19-273
Spaghetti, plain, fresh, cooked	see **Pasta, plain**		
Spare-ribs, pork, "barbecue style", chilled/frozen, reheated	see **Pork spare-ribs, "barbecue style"**		
Spare-ribs, pork, sliced, grilled, lean and fat	253		18-280
Speciality bread, white	see **Bread, speciality, white**		
Sponge cake, butter cream	63	11-422	
Sponge pudding, canned	78	11-396	11-328

	Publication number	New food code	Old food code
Sprats, raw	392		16-218
Spreadable butter	see **Butter, spreadable**		
Spread, blended, 40% fat	see **Blended spread, 40% fat**		
70-80% fat	see **Blended spread, 70-80% fat**		
Spread, fat, 20-25% fat, not polyunsaturated	see **Fat spread, 20-25% fat**		
35-40% fat, polyunsaturated	see **Fat spread, 35-40% fat**		
40% fat, not polyunsaturated	see **Fat spread, 40% fat**		
60% fat, with olive oil	see **Fat spread, 60% fat**		
70% fat, monounsaturated	see **Fat spread, 70% fat**		
70-80% fat, not polyunsaturated	see **Fat spread, 70-80% fat**		
Spring rolls, meat, takeaway	318	19-294	
Squid, raw	401		16-264
Starburst	see **Chew sweets**		
Steak and kidney pie, individual, cooked	319	19-290	19-069
Steak and kidney pudding, canned	320	19-291	19-072
Steak, braised, beef	see **Beef in sauce with vegetables**		
braising, beef, braised, lean and fat	see **Braising steak, beef**		
fillet, beef, cooked from steakhouse, lean only	see **Fillet steak, beef**		
rump, beef, fried in corn oil, lean only	see **Rump steak, beef**		
sirloin, beef, grilled, medium rare, lean only	see **Sirloin steak, beef**		
stewing, beef, stewed, lean and fat	see **Stewing steak, beef**		
Steaks, pork, grilled, lean only	see **Pork steaks**		
Stew, beef, meat only	see **Beef, stew**		
Irish, canned	see **Irish stew**		
Stewing lamb, stewed, lean and fat	238		18-187
Stewing steak, beef, stewed, lean and fat	220		18-081
Stilton, blue	135	12-867	12-180
Stir-fried beef with green peppers in black bean sauce	see **Beef, stir-fried**		
Stir-fried vegetable Thai curry, takeaway	428	15-355	
Stir-fried vegetables, takeaway	427	15-354	
Stock cubes, beef	510	17-478	17-368
Suet, shredded	169	17-473	17-011
Suet, vegetable	170		17-012
Sugar Puffs	29		11-152
Sun-dried tomatoes, in olive and sunflower oil	see **Tomatoes, sun-dried**		
Sunflower oil	202		17-045
Sunflower seeds	456	14-859	14-845
Sweet and sour pork, battered, takeaway	379	19-304	
Sweets, chew	see **Chew sweets**		
Swiss roll, chocolate	64	11-387	11-217
Szechuan prawns with vegetables, takeaway	411	16-320	
Tagliatelle with ham, mushrooms and cheese	380		19-279
Tagliatelle, plain, fresh, cooked	see **Pasta, plain**		

	Publication number	New food code	Old food code
Tandoori, chicken, chilled, reheated	see **Chicken tandoori**		
Taramasalata	413	16-313	16-307
Tart, Bakewell, individual	see **Bakewell tart**		
Tarts, custard, individual	see **Custard tarts**		
Tarts, jam, retail	see **Jam tarts**		
Taxi, chocolate wafer	see **Wafers, filled, chocolate**		
Tea whitener, powder	104	12-894	
Teacakes, retail	73	11-425	11-273
Thai curry, stir-fried vegetable, takeaway	see **Stir-fried vegetable Thai curry**		
Thousand island dressing	see **Dressing, thousand island**		
Tikka masala, chicken, takeaway	see **Chicken tikka masala**		
Tikka, chicken, chilled, reheated	see **Chicken tikka**		
Tip Top dessert topping	115	12-848	12-130
Toffee yogurt, low fat	see **Low fat yogurt, toffee**		
Toffees, mixed	481		17-120
Tofu dressing	see **Dressing, tofu**		
Tomato soup, cream of, canned	see **Canned cream of tomato soup**		
Tomato soup, cream of, dried	see **Dried cream of tomato soup**		
Tomatoes, sun-dried, in olive and sunflower oil	429		17-375
Tongue slices	356		19-154
Topside, beef, roasted, medium-rare, lean and fat	221		18-089
Torte, frozen/chilled, fruit	65	11-423	
Tortilla chips	498		17-149
Trex	see **Compound cooking fat**		
Trifle, chocolate, individual	158	12-878	12-252
Trifle, fruit	159	12-915	11-335
Trio bars	see **Chocolate biscuits, cream filled**		
Tripe, dressed, stewed	291	18-457	
Trout, rainbow, grilled, flesh only	394	16-315	16-226
Trout, rainbow, raw	393		16-225
Tuna, canned in brine, drained	395	16-316	16-229
Tuna, canned in oil, drained	396		16-230
Turkey roasts, frozen, cooked	357		19-155
Turkey sausages, raw	331	19-295	
Turkey, breast fillet, grilled, meat only	272		18-356
Turkey, dark meat, roasted	270	18-459	18-358
Turkey, light meat, roasted	271	18-460	18-359
Turkey, minced, stewed	273		18-354
Turkey, skin, raw,	269		18-351
Turkey, thighs, diced, casseroled,	274		18-355
Twiglets	499		17-150
Twin pot yogurt, virtually fat free/diet, fruit	see **Virtually fat free/diet yogurt**		
Twix	473	17-427	17-100
United bars	see **Chocolate biscuits, full coated**		

	Publication number	New food code	Old food code
Utterly Butterly	see **Fat spread, 70% fat, monounsaturated**		
Veal, escalope, fried in corn oil, lean	222		18-093
Veal, minced, stewed	223		18-095
Vegebanger mixes	433	15-349	15-323
Vegeburger mixes	432	15-348	15-332
Vegetable and cheese grill/burger, breaded, baked/grilled	435	15-359	
Vegetable balti, takeaway	430	15-356	
Vegetable biryani, takeaway	431	15-357	
Vegetable enchiladas, takeaway	see **Enchiladas, vegetable**		
Vegetable grill/burger, breaded, baked/grilled	434	15-358	
Vegetable kiev, baked	436	15-360	
Vegetable oil, blended	205	17-441	
Vegetable pâté	437	15-350	15-343
Vegetable samosa	438	15-351	15-305
Vegetable suet	see **Suet, vegetable**		
Vegetable Thai curry, stir-fried, takeaway	see **Stir-fried vegetable Thai curry**		
Vegetables, stir-fried, takeaway	see **Stir-fried vegetables**		
Vegetarian sausages, baked/grilled	439	15-361	
Venison, casseroled, meat only	283	18-455	
Venison, raw	282	18-467	18-390
Virtually fat free/diet yogurt, fruit, twin pot	143	12-907	
Vitalite	see **Fat spread, 70% fat, polyunsaturated**		
Vitalite Light	see **Fat spread, 35-40% fat, polyunsaturated**		
Wafers, Cornish	see **Cornish wafers**		
Wafers, filled, chocolate, full coated	53	11-416	
Waffles	see **Potato and tapioca snacks**		
Waffles, potato	see **Potato waffles**		
Walnut oil	203	17-439	17-047
Walnuts	457	14-860	14-850
Water biscuits	54	11-417	11-187
Weetabix	30	11-372	11-154
Wheat crunchies	500		17-151
Wheat flour, brown	3		11-028
Wheat flour, white, household, plain	4		11-031
Wheat flour, wholemeal	5		11-033
Wheatgerm oil	204	17-440	17-048
Whelks, boiled	402		16-268
White bread	12		11-099
White cap	see **Compound cooking fat**		
White rice, easy cook, raw	see **Rice, white**		
Whitener, coffee	see **Coffee whitener**		
tea, powder	see **Tea whitener**		
Whole milk yogurt, fruit	137	12-869	12-185

	Publication number	New food code	Old food code
Whole milk yogurt, fruit, infant	138	12-903	
Whole milk yogurt, plain	136	12-868	12-184
Whole milk, pasteurised, average	86	12-833	12-013
Whole milk, sterilised	87	12-834	12-017
Whole milk, UHT	88	12-835	12-016
Wholemeal bread, average	see **Bread, wholemeal**		
Wickettes	see **Potato and tapioca snacks**		
Winkles, boiled	403		16-270
Wispa bar	474	17-443	
Wotsits	see **Corn snacks**		
Yogurt coated nuts and raisins	445	14-862	
Yogurt, Greek style, plain	see **Greek style yogurt, plain**		
Yogurt, low fat	see **Low fat yogurt**		
Yogurt, reduced fat, frozen	see **Reduced fat yogurt, frozen**		
Yogurt, virtually fat free/diet, fruit, twin pot	see **Virtually fat free/diet yogurt**		
Yogurt, whole milk, fruit	see **Whole milk yogurt, fruit**		
plain	see **Whole milk yogurt, plain**